Numerical Analysis for Electromagnetic Integral Equations

For a listing of recent titles in the
Artech House Electromagnetic Analysis Series,
turn to the back of this book.

Numerical Analysis for Electromagnetic Integral Equations

Karl F. Warnick

ARTECH
HOUSE

BOSTON | LONDON
artechhouse.com

Library of Congress Cataloging-in-Publication Data
A catalog record for this book is available from the U.S. Library of Congress.

British Library Cataloguing in Publication Data
A catalogue record for this book is available from the British Library.

ISBN-13: 978-1-59693-333-0

Cover design by Yekaterina Ratner

© 2008 ARTECH HOUSE, INC.
685 Canton Street
Norwood, MA 02062

10 9 8 7 6 5 4 3 2 1

Contents

Preface

The subject of this book is error analysis of the moment method or boundary element method for solving surface integral equations of electromagnetic and acoustic radiation and scattering. With so many interesting applications of computational electromagnetics vying for attention, it can be difficult to devote time to the more academic issues associated with error analysis. The ultimate purpose of error analysis, though, is to advance the applications of computational electromagnetics, albeit by a less direct route than writing new solver codes.

When I first started publishing on the topic, paper introductions often included the thought that an increased understanding of numerical error will help to guide the development of new and improved algorithms. Chapter 5 is a good example of this hope coming to fruition. At a time when the computational electromagnetics community was studying and trying to improve magnetic field integral equation error using mostly empirical methods, error analysis provided a key insight that explained the typical poorer error with the magnetic field formulation and brought to light a method for increasing accuracy by orders of magnitude for simpler two-dimensional scatterers. A lesser but still encouraging improvement has been realized for three-dimensional problems.

To assist the reader in following the treatment of the book, a few main points should be underscored. Sobolev theory, which defines the domain and range spaces of integral operators, provides a theoretical foundation for the method of moments. This rigorous approach provided the first proofs of solution convergence—a significant achievement—but ultimately fails in providing quantitative error estimates. The purpose of this book is to obtain quantitative solution error estimates with enough analytical structure to understand the root causes of various error contributions. This is accomplished by relaxing the generality (and perhaps even some of the rigor) of the Sobolev approach by focusing on error analysis of canonical problems. Insights and results obtained using the canonical scatterers are extended empirically to more general classes of physical problems.

Error is quantified primarily through the surface current and scattering amplitudes. The latter is emphasized not because it is the only derived quantity that is useful in applications of the method of moments, but because the variational property of the scattering amplitude for the method of moments means that this particular quantity has a special relationship with the method of moments and its solution error behavior. Moreover, there is a connection between some Sobolev norms and the scattering amplitude.

One of the main themes of this book is that there are two sources of error in moment method solutions. The first is a multiplicative error associated with projection of a continuous current solution (the exact solution to a scattering problem) onto a finite and hence incomplete set of basis functions. We refer to this, naturally, as projection error. The second type of error is an additive "suboptimality" error contribution, which makes the moment method current solution less accurate than would be the case if the exact current solution were simply projected onto the basis functions. This error is caused by aliasing of high-order operator eigenfunctions. Since the eigenvalues associated with high-order eigenfunctions are determined by the kernel singularity, this aliasing error can be viewed as a projection error, but for which the integral operator kernel singularity (rather than the current solution) is projected onto the basis functions. From this point of view, solution error is determined by adding the errors incurred when projecting (1) the exact current solution and (2) the operator kernel onto basis functions. Interestingly, the current and scattering amplitude respond differently to these error contributions, which as will be shown is a manifestation of the variational property of the method of moments. For ideal discretizations, with conformal mesh elements and exact integration of moment matrix elements, the second source of error is minimized, but for nonideal implementations it increases significantly.

These error contributions will be analyzed by considering the spectral error, or the perturbation to operator eigenvalues incurred when discretizing an integral operator. The spectral error will facilitate the development of quantitative solution error bounds that closely match observed numerical results. The initial focus will be on simple, smooth two-dimensional scatterer geometries, and then effects such as edge current singularities and resonance will be considered. The treatment will be extended to three-dimensional scatterers and higher-order polynomial basis functions. Finally, the spectral estimates of earlier chapters will be used to study the convergence of iterative linear system solution algorithms and the dependence of moment matrix condition number on geometry and scattering physics.

I express particular appreciation to Weng Cho Chew, who initially suggested the error analysis problem to me in early 1998 during my postdoc at the University of Illinois at Urbana-Champaign. Professor Chew's emphasis on mathematical analysis coupled with physical insight helped my work immeasurably, and he has provided significant assistance with the writing and exposition of the ideas in this book. As mathematically tedious as some sections may be, they would be even more so without his encouragement to slow, simplify, and clarify the treatment. Clayton Davis significantly advanced the field of error analysis during his productive masters degree work at Brigham Young University, and many of his results are surveyed in this book. Andrew Peterson's long-standing interest in the error analysis problem helped to spur my own work in this area. His incisive comments and contributed numerical results have greatly added to this book as well. I also acknowledge Cai-Cheng Lu for generously allowing the use of his *trimom* method of moments code. I thank Barbara Lovenvirth and Rebecca Allendorf at Artech House and the anonymous reviewer for assistance with the production process. Finally, gratitude is expressed for the initial and subsequent insights supplied by divine providence that have enabled this work.

Chapter 1

Introduction

Numerical methods based on the surface integral equations of electromagnetic radiation and scattering have enjoyed widespread use in computational electromagnetics for many years. The method of moments for discretizing integrals equations was introduced to the electromagnetics community by Harrington [1]. Since that time, innovative techniques such as the fast multipole method [2] and iterative linear system solution algorithms have greatly increased the computational efficiency of surface integral equation methods. As the popularity and power of integral equation methods has increased, attention has also been devoted to characterizing the accuracy of numerical solutions and the development of higher-order methods and specialized basis functions, which brings us to the topic of this book, numerical analysis of the surface integral equations of electromagnetics.

Simply stated, numerical analysis is the science of developing numerical methods and understanding and improving their performance as far as solution accuracy and computational efficiency. An understanding of the convergence behavior of a numerical method requires an estimate or bound on the solution error in terms of parameters of the algorithm and physical properties of the problem to be solved. Error estimates will be developed for the integral equation formulations commonly used in computational electromagnetics, the electric field integral equation (EFIE), magnetic field integral equation (MFIE), and the combined formulation (CFIE). If the solution is obtained using an iterative method, then a complete understanding of the numerical method requires iteration convergence estimates as well. The intent is not just to provide solution error results, but also to connect solution accuracy and iterative convergence rates with electromagnetic effects and the physics of radiation and scattering problems, and to explain many of the behaviors of numerical methods that are commonly observed in computational electromagnetics.

As will be surveyed shortly, a fair amount of work has already been done by the mathematics community in this area. This work represents a significant achievement from a theoretical point of view but is often too abstract for direct use by practitioners of computational electromagnetics. A guiding philosophy for this book is to open up a new middle ground between the empirical focus of the computational electromagnetics community and the abstract theoretical approaches of numerical analysts.

1.1 APPROACHES TO ERROR ANALYSIS

It is well recognized in the computational electromagnetics community that error analysis for numerical methods based on integral equations is not easy. It has been claimed that "the nature of the MoM technique precludes proof of absolute convergence of the solution for a given problem" [3]. Only slightly less pessimistically, Dudley observed that [4]:

> Because the theory of nonself-adjoint operators is not well understood, it is not possible, based on any known mathematically sound convergence criteria, to compare the approximation obtained in the method of moments to the exact solution. This difficulty is typical of convergence problems in electromagnetics.

In spite of the difficulty of the problem, a large body of results on the solution convergence behaviors of numerical methods for electromagnetic integral equations has been amassed in recent years using both empirical and theoretical approaches.

Theoretical results on solution convergence are based on the theorems and techniques of functional analysis and operator theory. This work is rigorous, but is often too abstract for practical use and tends to lag behind the state-of-the-art in applied computational electromagnetics. Not content to wait for theory to catch up, users and developers of computational electromagnetics tools have used empirical means to validate their methods and allow practical work to proceed. Until now, the theoretical and empirical lines of work have proceeded largely independently. Both approaches have attractive features—one offering rigor and generality, and the other quantitative information useful to practitioners—but there are drawbacks as well. After surveying these areas of previous work, we will proceed to develop an approach to the numerical analysis of electromagnetic integral equations that attempts to combine the strengths of both of these lines of work.

1.2 EMPIRICAL METHODS

Empirical error analysis involves comparison of computed results with analytical solutions or measured data to determine the realized solution error. Empirical results are often combined with insights into the physical behavior of fields to predict or improve the accuracy and efficiency of a numerical method. This approach to methods development has been used almost exclusively in the computational electromagnetics community, and has been highly successful over the past two decades in driving research and applications in the field.

The most common validation case is the ubiquitous conducting sphere, although many other benchmark problems are available, including test case suites such as those maintained by the Electromagnetics Code Consortium (EMCC). Another avenue for empirical studies is cross-code validation of one numerical method using another type of numerical method, such as comparison of finite difference time domain (FDTD) and moment method results.

Although empirical verification can provide reasonable confidence of accuracy over classes of similar scatterer geometries, it yields only a limited understanding of the underlying causes of solution error and condition number growth. An important issue is the possible existence of problems for which error is large, despite good convergence for well-behaved test cases. The most common examples for which this difficulty arises are scatterers with sharp edges and resonant cavities or internal resonances.

To remedy the limitations of empirical validation, physical effects such as resonance and edge diffraction will receive careful study in this book. The analytical insights to be developed will provide a framework for understanding the large body of work on empirical validation that has been reported by the computational electromagnetics community.

1.3 Sobolev Spaces and Asymptotic Error Estimates

The theoretical work done by the numerical analysis community for electromagnetics problems [5–13] has its foundation in the classical study of Laplace's equation and the theory of Sobolev spaces. The techniques and tools of partial differential equation theory have been extended and applied to integral equations to achieve basic tasks such as proving that numerical solutions for a wide class of scatter geometries converge as the mesh is refined. As we will see shortly, this has been accomplished by obtaining bounds on the asymptotic solution convergence rate. Other key results include condition number bounds and the development of preconditioners that reduce the computational cost of solving large linear systems from the moment method to $O(1)$ computations per degree of freedom as the mesh density increases. Since Sobolev spaces provide the framework for most of the theoretical work done by the mathematics community on numerical analysis for electromagnetic integral equations, we will refer to this body of work as the Sobolev theory approach.

The governing parameter in an asymptotic error estimate is the mesh or discretization length h. For a triangular surface patch discretization, h represents the average or maximum edge length. Error estimates results obtained using the Sobolev space approach are asymptotic as $h \to 0$, and typically have the form

$$\|\hat{J} - J\|_{H^s} \leq ch^\alpha, \quad h < H \tag{1.1}$$

where c is an unknown constant, J represents the exact solution to the scattering problem, and $\hat{J}$ is the numerical solution obtained with the moment method. The parameter α is the rate or order of convergence of the numerical method and gives the log-log slope of the error as a function of the mesh refinement parameter h. Asymptotic convergence theorems guarantee that estimates of this type hold for h smaller than some unknown, problem-dependent constant H with a positive value for the convergence exponent α.

In this section, we will briefly sketch some of the main ideas and results in the Sobolev theory approach to moment method solution error analysis. This oversimplified discussion is not intended to be particularly rigorous, but instead to highlight the key results from this work and to connect Sobolev results with familiar concepts of electromagnetic theory.

1.3.1 Sobolev Norms

Both fields and currents lie in Sobolev spaces of various orders. Since we are interested in surface integral equations, we will focus on the Sobolev spaces associated with surface currents. A Sobolev space consists of all functions on a given domain that have a finite value for a particular Sobolev norm. For electromagnetic surface integral equations, the norm in (1.1) is a Sobolev norm associated with the space of possible surface current solutions for a given radiation or scattering problem. For functions defined on the real line, the Sobolev norm can be interpreted in the Fourier domain as a weighted norm, so that [14]

$$\|u\|_{H^s} = \int_\infty^\infty dk \, (1 + |k|)^{2s} |U(k)|^2 \tag{1.2}$$

where $U(k)$ is the Fourier transform of $u(x)$. The space of functions for which this norm is finite is denoted by H^s. For two-dimensional (2D) electromagnetic radiation and scattering problems with perfect electric conductor (PEC) bodies, it has been shown that surface currents lie in the fractional-order Sobolev spaces with order $s = \pm 1/2$. Surface currents induced on a cylindrical scatterer for the TM polarization belong to $H^{-1/2}$, and $H^{1/2}$ for the TE polarization.

For $s = 1/2$, the space $H^{1/2}$ is smaller than the square integrable function space L^2, and consists of functions smooth enough that the decay of $U(k)$ for large k offsets the growth of the weight function in the integrand of (1.2). For a scatterer with edges, the current solution for the TE polarization goes to zero at the edges. This means that the current solution is sufficiently smooth that (1.2) is finite. If the solution tended to a nonzero value at a scatterer edge, then the Fourier transform of the current would decay at a k^{-1} rate or slower, leading to an infinite value for the $H^{1/2}$ Sobolev norm.

For $s = -1/2$, the weight function acts as a smoothing operation, so functions that are less smooth than L^2 functions belong to the space $H^{-1/2}$. In particular, this Sobolev space contains functions with singularities of the form $x^{-1/2}$, $x \to 0$. For the TM polarization, surface currents can have $x^{-1/2}$ type singularities at edges [15]. The L^2 norm of such a current is infinite and cannot be used to define a meaningful current error, whereas the $H^{-1/2}$ norm remains finite and can be used to measure solution error in (1.1).

1.3.2 Sobolev Norms and the Scattering Amplitude

For computational electromagnetics practitioners, Sobolev spaces and norms have remained largely shrouded in mystery. The historical origins of Sobolev spaces, however, are quite well grounded in physics. The Sobolev space of electric fields and its associated norm merely make rigorous the observation that an electric field and its derivative (the associated magnetic field) must be square integrable in any bounded region [16]. The related Sobolev space for surface currents is the space of all possible functions on the surface that radiate square integrable fields.

In view of the physical basis for the Sobolev spaces of currents and fields, it is to be expected that the Sobolev norm in (1.1) also has physical meaning. In fact, there is a close relationship between the Sobolev norm and the scattering amplitude [17]. The one-dimensional Fourier transform of the 2D scalar Green's function has the same asymptotic behavior as the weight function in (1.2) as $|k| \to \infty$. For this reason, the Sobolev norm $\|J\|_{H^{-1/2}}$ is similar to the forward scattering amplitude $\langle E^i, \mathcal{L}J \rangle_{L^2}$, where E^i is an incident field and $\mathcal{L}J = E^i$ is the 2D electric field integral equation. The Sobolev norm can be viewed as a forward scattering amplitude that is modified to satisfy the properties required for a norm, essentially by taking the absolute value of the Fourier transform of the 2D Green's function and removing its singularities at $k = k_0$, where k_0 is the wavenumber of the incident field.

As a corollary to this result, a bound on the Sobolev norm of the current error implies a bound on the error of the numerically computed backscattering amplitude, so (1.1) for $s = -1/2$ can be thought of as an error bound for the backscattering amplitude. This equivalence has been shown to hold for 3D radiation and scattering problems as well [17]. In view of this connection between Sobolev norms and the scattering amplitude, a major focus in this book will be on the scattering amplitude solution error as a measure for the accuracy of the method of moments.

1.3.3 Static Limit

A basic result that can be used to obtain solution error bounds of the form (1.1) is that the convergence behavior of MoM for a dynamic problem for small h is close in a certain sense to that of the static or Laplace problem. This reduces the error analysis from that of the nonself-adjoint integral operators of electromagnetics for dynamic problems to the simpler integral operator associated with the Laplace problem. From the operator point of view, this amounts to decomposing the integral operator in the form

$$\mathcal{L} = (j/k_0)\mathcal{A} + \mathcal{K} \tag{1.3}$$

where the static part $\mathcal{A}$ is self-adjoint and positive definite and $\mathcal{K}$ is compact, and neglecting the influence of $\mathcal{K}$ on the convergence behavior of the method of moments.

1.3.4 Quasioptimality and Approximation Error

Another fundamental principle in the theory of error analysis is quasioptimality. In some cases, the static approximation in (1.3) can be used to show that the moment method solution for a given surface integral equation is close to the best possible solution in the trial subspace, with the notion of closeness defined in terms of a Sobolev norm. As a consequence, the actual error is not significantly different from the approximation error, which is the error that would be obtained with the best possible combination of basis functions.

Quasioptimality leads to a drastic simplification in the error analysis problem, since it reduces error analysis to a problem in approximation theory, and the original electromagnetic integral equation is completely out of the picture. Details of the implementation of the numerical algorithm become irrelevant and all that matters is the smoothness properties of the exact current solution and how well the solution is represented by the basis functions. The ability of the basis functions to represent functions of the given smoothness class then determines a solution error estimate of the form (1.1).

1.3.5 Convergence Theorems

The order of convergence α in (1.1) can be evaluated for smooth scatterers and scatterers with edges. For smooth scatterers, the current is continuous and differentiable everywhere, as long as there are no sources on the scatterer itself. The typical value for the convergence rate obtained using the Sobolev approach is $\alpha = 1/2$, for 2D smooth closed curves and screens [18], dielectric polygons [19], and nonsmooth geometries [7]. Similar estimates are available for smooth 3D scatterers with scalar (acoustic) fields [20] and 3D vector fields [5, 6].

Another application of Sobolev theory is the development of a posteriori residual-based error bounds [14, 21, 22]. These bounds give the solution error in terms of the residual error, or error in the scattered field, and require accurate computation of residuals in order to bound the solution error. One application of a posteriori error bounds is error indicators for adaptive grid refinement [9].

The assumptions required in Sobolev theory proofs of asymptotic error estimates as far as scatterer geometry and other physical properties of a scattering problem are quite weak, so that error estimates are valid for essentially any scattering problem. Because of this high degree of generality, the asymptotic error estimates provide an important result—that numerical solutions converge to the correct answer for a wide range of radiation and scattering problems. Until these theorems were obtained, no assurance other than empirical observations was available that numerical methods for the integral equations of radiation and scattering are well founded. Unfortunately, observed solution convergence rates are much faster than $h^{1/2}$, so these bounds only indicate that the solution converges and provide no further quantitative information. This is one of the major limitations of the Sobolev theory approach to error analysis to be discussed in the next section.

1.3.6 Limitations of Existing Error Estimates

The price paid for the generality of the Sobolev theory-based results includes the lack of tightness in the convergence rate α, the presence of unknown constants in the error estimate (1.1), and the difficulty of connecting the Sobolev norm of the solution error to error in physical parameters. For practical situations where quantitative information on the actual solution error for a given scattering problem is desired, the asymptotic error estimates obtained from Sobolev theory are inadequate.

In applications of computational electromagnetics, it would be desirable to have error estimates that improve on the Sobolev theory results in a number of ways:

Simple error norms: Since the Sobolev norm is difficult to compute, and even if it could be computed, it may not be clear how to interpret the norm in terms of physical quantities, estimates for simple error measures such as discrete RMS current error, error in the continuous L^2 norm, or scattering amplitude error are needed.

Tight convergence orders: Theoretical predictions for the asymptotic solution convergence order α should match empirical observations, and should reflect the influence of scatterer geometry and basis functions on solution accuracy.

Known error constants: For practical applications, error estimates with known values for the constant c would be beneficial, so that quantitative error bars for numerical solutions can be given.

Characterization of the asymptotic convergence regime: In (1.1), the constant H that determines the asymptotic convergence regime $h < H$ is unknown. The on-set of asymptotic convergence depends on the frequency and geometrical shape of the scatterer, as well as details of the numerical method. Physical effects such as resonance may cause H to become very small, so that the asymptotic regime is not achieved for computationally feasible meshes with reasonable numbers of unknowns.

1.4 Spectral Convergence Theory

The intent of this book is to address the drawbacks of earlier work and to provide a practical understanding of error and convergence behavior for the electrically large, complex radiation and scattering problems of current interest. A more concrete error analysis approach is needed that explicitly includes the effect of wave physics in a description of solution accuracy and numerical behavior. This will be accomplished by developing an error theory based on a spectral description of the integral operators of radiation and scattering.

1.4.1 Normal Operator Decomposition

Most classical theorems of error analysis apply to self-adjoint operators. For static Laplace boundary value problems, integral operators are self-adjoint, but for dynamic radiation and scattering problems, integral operators associated with Helmholtz and Maxwell boundary value problems are nonself-adjoint and are nonnormal as well except in a few special cases. This observation motivates the decomposition (1.3), which allows theorems for the self-adjoint case to be applied to dynamic operators, subject to the condition that the perturbation is small. Dynamic effects associated with resonance and large electrical size can have a significant effect on solution accuracy, which accounts for some of the limitations of asymptotic error estimates described in Section 1.3.6.

In order to improve upon existing asymptotic estimates, the spectral approach to error analysis is based on decompositions of the form

$$\mathcal{L} = \mathcal{H} + \mathcal{R} \tag{1.4}$$

where $\mathcal{H}$ is normal rather than self-adjoint as in (1.3) and has an exactly or approximately known discrete spectrum, and $\mathcal{R}$ is a nonnormal perturbation. The normal operator will be used to provide estimates of the eigenvalues of the integral operator.

1.4.2 Spectral Error

Physical properties of the scattering problem can be parameterized through their effects on the spectrum of the operator as estimated from $\mathcal{H}$. The discretization process used to transform an integral equation into a finite linear system can be treated as an effective truncation and perturbation of the spectrum. This allows the effects of discretization, quadrature error, and other aspects of a numerical algorithm on the operator spectrum to be quantified, and provides spectral estimates from which solution error and condition number estimates can be obtained.

We will refer to the perturbation to the operator eigenvalues caused by discretization as spectral error. This error will be analyzed by representing the spectrum of the discretized operator or moment matrix in terms of the eigenvalues of the continuous operator $\mathcal{H}$. The spectral error theory will lead to several basic concepts that will appear repeatedly throughout the book:

> *Eigenfunctions and modes:* The surface integral operators of electromagnetic radiation and scattering have a countably infinite number of eigenfunctions. In many cases, these eigenfunctions can be approximated by Fourier type oscillatory modes with small perturbations due to such effects as edge diffraction. For some canonical scatterers, such as the circular cylinder or sphere, the eigenfunctions are exactly equal to Fourier modes or spherical harmonics.

> *Mesh Nyquist frequency:* Discretization of a continuous current on a mesh imposes an upper bound on the spatial frequencies that can be represented. We refer to this bound as the mesh Nyquist frequency.

> *Modeled modes:* Modeled modes have spatial frequency below the mesh Nyquist frequency, and can be well represented in terms of basis functions defined on the mesh. The number of modeled modes is equal to the size of the moment matrix.

> *Unmodeled modes:* Unmodeled modes have spatial frequency above the mesh Nyquist frequency and cannot be accurately represented on the mesh. The asymptotic decay rate of the operator eigenvalues associated with unmodeled modes with the mode order or spatial frequency is determined by the kernel singularity of the integral operator. If the kernel is nonsingular, the eigenvalues of unmodeled modes are very small. For singular kernels such as the scalar Green's function, the eigenvalues are larger and decay slowly with order. For integral operators that include derivatives, the eigenvalues can grow as the order becomes large.

> *Spectral error:* The eigenvalues of the moment matrix are numerically close in value to the eigenvalues of the continuous integral operator for the modeled modes. The difference between the eigenvalues of the moment matrix and the corresponding continuous operator eigenvalues for the modeled modes is spectral error.

> *Projection error:* The spectral error divides naturally into two contributions, one that is a scale factor associated with projection of a modeled mode onto a space of basis functions, and another additive component that is associated with the eigenvalues of unmodeled modes. The first of these contributions is projection error.

> *Aliasing error:* When discretized, the unmodeled modes are above the mesh Nyquist frequency and so they alias to lower order, modeled modes. As a re-

sult, the eigenvalues of unmodeled modes add to the eigenvalues of modeled modes and contribute to the spectral error. Physically, aliasing error is due to quasistatic radiation from the basis functions themselves, and edges or corners for low order basis functions cause larger aliasing error, whereas smoother basis functions reduce it.

Of course, the real goal is not an understanding of the spectral error itself. We want to characterize errors for measurable physical quantities such as the surface current, radiated fields, scattering amplitudes, and radar cross section. After analyzing spectral error, relationships are developed between the spectral error and the error in numerical values computed from the method of moments for these physical quantities. One of the most interesting results will be that currents and scattering amplitudes respond differently to spectral error. For ideal discretizations, with conformal mesh elements and exact integration of moment matrix elements, the scattering amplitude is only sensitive to the additive part of the spectral error due to aliasing of unmodeled modes. It will be demonstrated that this is closely related to the well-known variational property of the method of moments.

1.4.3 Error Contributions

Solution error is determined by interactions between the physical properties of the scattering problem and the numerical method used to solve it. Specific properties of the problem, mesh representation, and the numerical method that affect the accuracy of the method of moments are:

Problem

> Scatterer smoothness—edge, corner, and point singularities.
>
> Geometry of the scatterer—internal resonances and open-cavity resonances.
>
> Incident electromagnetic field or excitation—angle of incidence and type of source (plane wave or near field source such as a dipole).
>
> Type of final desired numerical result—current, scattering cross-section, total field, or scattered field.

Mesh

> Mesh density (elements per wavelength).
>
> Low-frequency breakdown for electrically small scatterers.
>
> Element size irregularity and mesh defects.
>
> Geometrical discretization error—flat or curved facets.

Numerical Method

Integral equation formulation—EFIE, MFIE, or CFIE.

Expansion and testing functions—type of basis and polynomial order.

Quadrature rule used to evaluate moment matrix elements.

Linear system solution algorithm—direct factorization or iterative.

Some of these factors, such as smoothness of the scatterer geometry and the choice of basis functions, influence the convergence rate of the solution. Other factors, such as electrical size and resonance, impact the absolute accuracy of the method but do not affect the decay rate of the error as a mesh is refined. Both types of error contributions will be considered using the tools of spectral convergence theory.

1.4.4 Canonical Scattering Problems

We will consider each of the error factors listed above by applying the spectral error approach to several canonical scattering examples. We also show how error contributions can be superimposed to estimate the error for more complex scatterers. To provide a simple example with which to develop the concepts of spectral error analysis, we will first consider the circular cylinder. Since the scattering problem for a circular cylinder has an analytical series solution, the spectral error approach can be applied directly, without the normal approximation of (1.4), so the treatment is particularly straightforward. We will consider ideal discretizations, with conformal mesh elements and exact integration of moment matrix elements, and then add geometrical discretization error and quadrature error to the analysis. Methods for improving solution accuracy such as regularization will be analyzed using these tools.

To extend the spectral error analysis to more complex problems, we will consider the flat strip as a scatterer with edge singularities. Error due to currents on the smooth part of the strip away from the edge singularities can be combined with error caused by edge diffraction effects to obtain a total error estimate for the strip. Internal and real resonances also affect the solution convergence rate. The cylinder exhibits internal resonances, which are nonphysical but still influence numerical accuracy. The rectangular cavity is an example of a scatterer with real, physical resonances. Spectral error estimates for these scatterers allow the error due to resonance effects to be quantified.

In order to apply the spectral error analysis approach to the more general 3D case, we study the behavior of the method of moments for a flat, conducting plate. Higher order basis functions for 2D and 3D problems are then considered. Finally, we consider the use of iterative linear system solution algorithms and study the condition number of the method of moments for 2D and 3D scatterers, including resonant and nonresonant cases.

The main contributions of this book are quantitative solution error bounds that closely match computed results and a theory that explains a wide range of phenomena observed in computational electromagnetics, including mesh element size dependence of error, error for scattered with edges and corners, poor accuracy at internal resonance frequencies, cavity resonance effects, and the differences in errors among the various integral formulations (EFIE, MFIE, and CFIE). These results will provide a theoretical understanding of observed numerical behaviors and a sound basis for extrapolating empirical observations for canonical geometries to more general problems of practical interest, thereby bridging the gap between empirical error studies of the computational electromagnetics community and the abstract numerical analysis of the mathematics community.

References

[1] R. F. Harrington, *Field Computation by Moment Method*. Malabar, FL: Krieger, 1982.

[2] C. C. Lu and W. C. Chew, "A multilevel algorithm for solving a boundary integral equation of scattering," *Micro. Opt. Tech. Lett.*, vol. 7, pp. 466–470, July 1994.

[3] M. B. Gedera, L. N. Medgyesi-Mitschang, R. Pearlman, J. M. Putnam, and D. Wang, "Modeling accuracy of method of moments," *11th Annual Review of Progress in Applied Computational Electromagnetics*, vol. II, Monterey, CA, pp. 1194–1201, Naval Postgraduate School, Mar. 20–25, 1995.

[4] D. G. Dudley, "Error minimization and convergence in numerical methods," *Electromagnetics*, vol. 5, pp. 89–97, 1985.

[5] L. Demkowicz, "Asymptotic convergence in finite and boundary element methods: Part 1: Theoretical results," *Computers Math. Applic.*, vol. 27, no. 12, pp. 69–84, 1994.

[6] H. Holm and E. P. Stephan, "A boundary element method for electromagnetic transmission problems," *Appl. Anal.*, vol. 56, pp. 213–226, 1995.

[7] H. Holm, M. Maischak, and E. P. Stephan, "The hp-version of the boundary element method for Helmholtz screen problems," *Computing*, vol. 57, pp. 105–134, 1996.

[8] M. Ainsworth, W. McLean, and T. Tran, "The conditioning of boundary element equations on locally refined meshes and preconditioning by diagonal scaling," *SIAM J. Numer. Anal.*, vol. 36, no. 6, pp. 1901–1932, 1999.

[9] M. Maischak, P. Mund, and E. P. Stephan, "Adaptive multilevel BEM for acoustic scattering," *Comp. Meth. Appl. Mech. Engrg.*, vol. 150, pp. 351–367, 1997.

[10] T. Tran, E. P. Stephan, and P. Mund, "Hierarchical basis preconditioners for first kind integral equations," *Appl. Anal.*, vol. 65, pp. 353–372, 1997.

[11] S. Amini and N. D. Maines, "Preconditioned Krylov subspace methods for boundary element solution of the Helmholtz equation," *Int. J. Num. Meth. Engr.*, vol. 41, pp. 875–898, 1998.

[12] E. P. Stephan and T. Tran, "Domain decomposition algorithms for indefinite hypersingular integral equations: The h and p versions," *SIAM J. Sci. Comput.*, vol. 19, pp. 1139–1153, July 1998.

[13] E. P. Stephan and T. Tran, "Domain decomposition algorithms for indefinite weakly singular integral equations: The h and p versions," *IMA J. Numer. Anal.*, vol. 20, no. 1, pp. 1–24, 2000.

[14] E. F. Kuester, "Computable error bounds for variational functionals of solutions of a convolution integral equations of the first kind," *Wave Motion*, vol. 22, pp. 171–185, 1995.

[15] J. Van Bladel, *Singular Electromagnetic Fields and Sources*. New York: IEEE Press, 1995.

[16] G. C. Hsiao and R. E. Kleinman, "Mathematical foundations for error estimation in numerical solutions of integral equation in electromagnetics," *IEEE Trans. Ant. Propag.*, vol. 45, pp. 316–328, Mar. 1997.

[17] C. P. Davis and K. F. Warnick, "On the physical meaning of the Sobolev norm in error estimation," *Journal of the Applied Computational Electromagnetics Society*, vol. 20, pp. 144–150, July 2005.

[18] M. Feistauer, G. C. Hsiao, and R. E. Kleinman, "Asymptotic and a posteriori error estimates for boundary element solutions of hypersingular integral equations," *SIAM J. Numer. Anal.*, vol. 33, pp. 666–685, Apr. 1996.

[19] M. Costabel and E. P. Stephan, "A direct boundary integral equation method for transmission problems," *J. Math. Anal. Appl.*, vol. 106, pp. 367–413, 1985.

[20] E. P. Stephan, "Boundary integral-equations for screen problems in IR-3," *Integral Equations and Operator Theory*, vol. 10, pp. 236–257, 1987.

[21] M. Feistauer, G. C. Hsiao, and R. E. Kleinman, "Asymptotic and a posteriori error estimates for boundary element solutions of hypersingular integral equations," *SIAM J. Numer. Anal.*, vol. 33, pp. 666–685, Apr. 1996.

[22] G. C. Hsiao and R. E. Kleinman, "Feasible error estimates in boundary element methods," in *Boundary Element Technology VII* (C. A. Brebbia and M. S. Ingber, eds.), pp. 875–886, Southampton: Computational Mechanics Publ., 1991.

Chapter 2

Surface Integral Equation Formulations and the Method of Moments

For conducting or dielectric bodies, Maxwell's equations and the boundary conditions on the electromagnetic fields at the surface of the scatterer can be cast into an equivalent system of surface integral equations. These equations are based on integral operators that relate an unknown, equivalent current on the surface of the body to the fields scattered in response to a given incident field. We consider only perfectly electrically conducting (PEC) scatterers, although the results of this book could be readily extended to surface integral formulations for homogeneous dielectric objects. We will also leave untouched the problem of error analysis for volume integral equations.

We will use a single-frequency (time-harmonic) analysis, so that all field quantities are represented as phasors, with the real field related to the phasor representation according to

$$\mathcal{E}(\mathbf{r}, t) = \operatorname{Re}\left\{\mathbf{E}(\mathbf{r})e^{j\omega t}\right\} \tag{2.1}$$

for the electric field intensity, where $\mathbf{r}$ represents a point (x, y, z) and ω is the time frequency of the field in rad/sec. We will work exclusively with the complex field quantity $\mathbf{E}$, which is the phasor representation of the time-harmonic field. Phasors associated with the magnetic field are defined similarly. Another common convention for phasors uses a negative sign for the exponent and i for the imaginary unit. The convention used in this book can be converted to the $e^{-i\omega t}$ form by taking complex conjugates of field quantities.

The model physical problem with which we are concerned is a PEC scatterer illuminated by an incident wave $\mathbf{E}^{\mathrm{inc}}$. The surface of the PEC object is denoted by S. The incident wave induces a surface current $\mathbf{J}_s$ on the scatterer, which radiates a scattered field $\mathbf{E}^{\mathrm{sca}}$. The total field is $\mathbf{E} = \mathbf{E}^{\mathrm{inc}} + \mathbf{E}^{\mathrm{sca}}$. At the surface of the PEC scatterer, the boundary conditions

$$\hat{n} \times \left(\mathbf{E}^{\text{inc}} + \mathbf{E}^{\text{sca}} \right) = 0 \tag{2.2a}$$

$$\hat{n} \times \left(\mathbf{H}^{\text{inc}} + \mathbf{H}^{\text{sca}} \right) = \mathbf{J}_s \tag{2.2b}$$

must be satisfied, where $\hat{n}$ is a unit surface normal vector. For a PEC scatterer, only one boundary condition suffices for the solution. Using the radiation integral and free-space Green's function, we can relate the scattered fields $\mathbf{E}^{\text{sca}}$ and $\mathbf{H}^{\text{sca}}$ to the induced surface current $\mathbf{J}_s$. Combining these integral representations with either of the boundary conditions leads to an equation in which the incident field is given and the surface current is unknown. Since the unknown current appears under an integral operator, the relationships are integral equations.

Most commonly, the incident field is taken to be a homogeneous plane wave of the form

$$\mathbf{E}^{\text{inc}}(\mathbf{r}) = \mathbf{E}_0 e^{-j\mathbf{k}\cdot\mathbf{r}} \tag{2.3}$$

where $\mathbf{E}_0$ is a constant vector that determines the amplitude and polarization of the field and the vector $\mathbf{k}$ determines the direction of propagation. The magnitude $k_0 = |\mathbf{k}| = \sqrt{k_x^2 + k_y^2 + k_z^2}$ is the wavenumber of the incident field, and is related to the properties of the medium surrounding the scatterer according to the dispersion relation $k_0 = \omega\sqrt{\mu\epsilon}$. For a radiation problem, the incident field is generated by a source such as an antenna feed port near to the scatterer S, and $\mathbf{E}^{\text{inc}}$ is not a plane wave.

2.1 ELECTRIC FIELD INTEGRAL EQUATION

When the boundary condition (2.2a) is imposed, one needs to first find $\mathbf{E}^{\text{sca}}$, which can be related to the surface current $\mathbf{J}_s$ using

$$\mathbf{E}^{\text{sca}} = -\mathcal{T}\mathbf{J}_s \tag{2.4}$$

where the integral operator is [1, 2]

$$\mathcal{T}\mathbf{J}_s = jk_0\eta \int_S ds'\, g(\mathbf{r}, \mathbf{r}')\mathbf{J}_s(\mathbf{r}') + \frac{j\eta}{k_0} \nabla \int_S ds'\, g(\mathbf{r}, \mathbf{r}')\nabla' \cdot \mathbf{J}_s(\mathbf{r}') \tag{2.5}$$

This definition of the operator $\mathcal{T}$ differs from that of [2] in that it does not include the cross product with the surface normal vector $\hat{n}$. In this expression, $\eta = \sqrt{\mu/\epsilon}$ is the characteristic impedance of the medium surrounding the scatterer, ϵ and μ are the permittivity and permeability of the medium, k_0 is the wavenumber of the time harmonic incident field, and the kernel is the free space Green's function

$$g(\mathbf{r}, \mathbf{r}') = \frac{e^{-jk_0 R}}{4\pi R} \tag{2.6}$$

where $R = |\mathbf{r} - \mathbf{r}'|$.

Consequently, the electric field integral equation (EFIE) is of the form

$$\hat{n} \times \mathcal{T} \mathbf{J}_s = \hat{n} \times \mathbf{E}^{\text{inc}} \tag{2.7}$$

The region of integration in (2.5) is a two-dimensional manifold or surface S lying in three-dimensional space. To impose the boundary condition, the source point $\mathbf{r}'$ ranges over S, where the current $\mathbf{J}_s$ is nonzero. The field point $\mathbf{r}$ is also located on S. For the EFIE, the surface may be closed, so that S is the boundary of a volumetric region, or open, in which case S is a sheet with edges, corresponding to a thin conducting screen.

2.1.1 2D Scattering Problems

For a translationally invariant or infinite cylindrical scatterer, the three-dimensional scattering problem reduces to a pair of independent, scalar, two-dimensional Helmholtz boundary value problems. By convention we take z to be the invariant direction. If the polarization of the incident electric field is in the invariant direction, the magnetic field is then transverse to z, and the problem is transverse magnetic (TM). The EFIE reduces to a scalar integral equation of the form

$$\mathcal{L} J_z = E_z^{\text{inc}} \tag{2.8}$$

where the TM-EFIE integral operator is [3]

$$\mathcal{L} J_z = \frac{k_0 \eta}{4} \int_C ds' \, H_0^{(2)}(k_0 |\boldsymbol{\rho} - \boldsymbol{\rho}'|) J_z(\boldsymbol{\rho}') \tag{2.9}$$

For 2D problems, C denotes a path in the plane determined by the longitudinal cross-section of the 3D scatterer surface S. C may be a closed contour or an open arc.

For the transverse electric (TE) polarization, the magnetic field is in the z direction. The TE-EFIE is

$$\mathcal{N} J_t = E_t^{\text{inc}} \tag{2.10}$$

where J_t is the tangential component of the surface current density vector relative to the contour C and E_t^{inc} is the tangential component of the incident electric field. The integral operator is [3]

$$\begin{aligned} \mathcal{N} J_t = \frac{k_0 \eta}{4} \hat{t} \cdot \int_C ds' \, H_0^{(2)}(k_0 |\boldsymbol{\rho} - \boldsymbol{\rho}'|) \hat{t}' J_t(\boldsymbol{\rho}') \\ + \frac{1}{4 k_0 \eta} \hat{t} \cdot \nabla \int_C ds' \, H_0^{(2)}(k_0 |\boldsymbol{\rho} - \boldsymbol{\rho}'|) \nabla' \cdot [\hat{t}' J_t(\boldsymbol{\rho}')] \end{aligned} \tag{2.11}$$

where $\hat{t}$ is a unit vector tangential to C and transverse to the invariant direction.

The operator $\mathcal{L}$ is weakly singular, because the logarithmic singularity of $H_0^{(2)}(x)$ as $x \to 0$ is weak and relatively easy to integrate numerically. The operator $\mathcal{N}$ is hypersingular, because the derivatives in the second term of (2.11) increase the singularity of the kernel of the integral operator. In practice, the stronger singularity of the operator $\mathcal{N}$ increases the difficulty of evaluating the numerical integrations required to discretize the EFIE for the TE polarization.

2.2 MAGNETIC FIELD INTEGRAL EQUATION

The magnetic field integral equation (MFIE) is obtained by imposing the boundary condition (2.2b) to obtain

$$\mathbf{J}_s = -\mathcal{K}\mathbf{J}_s + \hat{n} \times \mathbf{H}^{\mathrm{inc}} \tag{2.12}$$

where the integral operator is [3]

$$\mathcal{K}\mathbf{J}_s = -\hat{n} \times \int_S ds' \nabla g(\mathbf{r}, \mathbf{r}') \times \mathbf{J}_s(\mathbf{r}') \tag{2.13}$$

and the point $\mathbf{r}$ lies just outside the scatterer S. The above integral equation can be manipulated into different forms. The integral on the right-hand side of (2.13) is strongly singular as the point $\mathbf{r}$ approaches $\mathbf{r}'$ in a direction normal to the surface S. In the limit as the point $\mathbf{r}$ approaches S from the outside, the principal value of the integral must be augmented according to [1]

$$\lim_{\mathbf{r} \to S} \hat{n} \times \int_S ds' \nabla g(\mathbf{r}, \mathbf{r}') \times \mathbf{J}_s(\mathbf{r}') = \tfrac{1}{2}\mathbf{J}_s - \mathcal{K}_p\mathbf{J}_s \tag{2.14}$$

where

$$\mathcal{K}_p\mathbf{J}_s = -\hat{n} \times \mathrm{PV} \int_S ds' \nabla g(\mathbf{r}, \mathbf{r}') \times \mathbf{J}_s(\mathbf{r}') \tag{2.15}$$

and $\mathbf{r}$ is assumed to approach the surface S from outside the scatterer. The integral in (2.15) must be interpreted as a principal value integration, although this is typically not explicitly represented in the notation for the MFIE. Consequently, the MFIE is

$$\tfrac{1}{2}\mathbf{J}_s + \mathcal{K}_p\mathbf{J}_s = \hat{n} \times \mathbf{H}^{\mathrm{inc}} \tag{2.16}$$

In the following, we will drop the subscript indicating the principal value and use the notation $\mathcal{K}$ for the integral operator.

Alternatively, the boundary condition (2.2b) can be written as

$$\hat{n} \times \left[\mathbf{H}^{\mathrm{inc}}(\mathbf{r}) + \mathbf{H}^{\mathrm{sca}}(\mathbf{r}) \right] = 0 \tag{2.17}$$

where the field point $\mathbf{r}$ approached the surface S from inside the scatterer. In other words, the incident field and the scattered field cancel each other exactly inside the scatterer (this is also known as the extinction theorem [1]). When $\mathbf{r}$ approaches the surface from inside the scatterer, the first term in (2.14) will assume a negative sign. When the expression for the scattered field is substituted into (2.17), we obtain the same integral equation as before.

Because of the leading term $\frac{1}{2}\mathbf{J}_s$, the MFIE is a second-kind integral equation, and the operator can be viewed as a perturbation of the identity operator. If the integral operator in (2.16) is neglected, the MFIE reduces to the physical optics approximation $\mathbf{J}_s \simeq 2\hat{n} \times \mathbf{H}^{\text{inc}}$. For the MFIE, the scatterer S must be a closed surface.

2.2.1 2D Scattering Problems

For 2D problems with a TM polarized incident field, the TM-MFIE has the form

$$\tfrac{1}{2}J_z + \mathcal{M}_{\text{TM}}J_z = H_t^{\text{inc}} \tag{2.18}$$

where H_t^{inc} is the tangential component of the incident magnetic field. The integral operator is

$$\mathcal{M}_{\text{TM}}J_z = \frac{jk_0}{4} \int_S ds' \, \cos\psi H_1^{(2)}(k_0|\boldsymbol{\rho} - \boldsymbol{\rho}'|)J_z(\boldsymbol{\rho}') \tag{2.19}$$

where J_z represents the longitudinal component of the current density and ψ is the angle between the surface normal at $\boldsymbol{\rho}$ and the vector $\boldsymbol{\rho} - \boldsymbol{\rho}'$.

For the TE polarization, the tangential current J_t is transverse to z, and flows tangentially on the boundary of the scatterer. The TE-MFIE integral equation is

$$\tfrac{1}{2}J_t + \mathcal{M}_{\text{TE}}J_t = H_z^{\text{inc}} \tag{2.20}$$

where the operator is

$$\mathcal{M}_{\text{TE}}J_t = \frac{jk_0}{4} \int_S ds' \, \cos\psi' H_1^{(2)}(k_0|\boldsymbol{\rho} - \boldsymbol{\rho}'|)J_t(\boldsymbol{\rho}') \tag{2.21}$$

and ψ' is the angle between the surface normal at $\boldsymbol{\rho}'$ and the vector $\boldsymbol{\rho} - \boldsymbol{\rho}'$.

2.3 COMBINED FIELD INTEGRAL EQUATION

A difficulty with the MFIE and EFIE is that for closed scatterers, at certain frequencies the integral operators $\mathcal{T}$ or $\frac{1}{2} + \mathcal{K}$ become singular. These correspond to nontrivial solutions to the interior Maxwell boundary value problem, or internal resonances of the cavity formed by the interior of the scatterer S. For the EFIE, the operator becomes singular at internal resonances with a PEC boundary condition, whereas for the MFIE the internal resonances are associated with a perfect magnetic conductor (PMC) boundary condition. At internal resonance frequencies, numerical solutions typically become inaccurate.

There are a number of techniques for overcoming the difficulties associated with internal resonances. The most common is to discretize a linear combination of the MFIE and EFIE. This is the combined field integral equation formulation (CFIE). The EFIE and MFIE have vanishing eigenvalues at certain real frequencies, but the two operators can be combined in such a way that the eigenvalues of the CFIE remain nonzero for all real frequencies.

The CFIE can be written in the form

$$\left[\alpha\hat{n} \times \mathcal{T} + (1 - \alpha)\eta\left(\tfrac{1}{2} + \mathcal{K}\right)\right]\mathbf{J} = \alpha\hat{n} \times \mathbf{E}^{\text{inc}} + (1 - \alpha)\eta\hat{n} \times \mathbf{H}^{\text{inc}} \qquad (2.22)$$

where α is a combination coefficient in the interval $[0, 1]$ and $\mathcal{K}$ has a principal value interpretation as described in Section 2.2. The factor of η is included so that both terms have the same physical units.

It can be shown that CFIE reduces the internal resonance solution to that of a lossy cavity whose resonance solutions have complex frequencies. Hence, these resonance solutions are never encountered when one seeks a time-harmonic solution for which the frequency is real [4, 5].

The combination coefficient α is commonly taken to be 0.2, which has been shown to minimize current error under some conditions for 2D problems [6]. Helaly and Fahmy [7] cite [8] as having recommended this value in 1970. An optimal specification for the coupling parameter is studied for the 3D case in [9].

2.3.1 2D Scattering Problems

For 2D scattering problems, the CFIE is

$$\left[\alpha\mathcal{L} + (1 - \alpha)\eta(\tfrac{1}{2} + \mathcal{M}_{\text{TM}})\right]J_z = \alpha E_z^{\text{inc}} + (1 - \alpha)\eta H_t^{\text{inc}} \qquad (2.23)$$

for the TM polarization. For the TE polarization,

$$\left[\alpha\mathcal{N} + (1 - \alpha)\eta(\tfrac{1}{2} + \mathcal{M}_{\text{TE}})\right]J_t = \alpha E_t^{\text{inc}} + (1 - \alpha)\eta H_z^{\text{inc}} \qquad (2.24)$$

2.4 METHOD OF MOMENTS

The surface integral equations of radiation and scattering must be discretized by projecting into finite-dimensional subspaces in order to be solved numerically. Discretization yields a linear system of equations for samples or weights that determine an approximation to the unknown surface current on the scatterer.

For the scalar or 2D case, the surface current density is approximated by an expansion of the form

$$J(s) \simeq \hat{J}(s) = \sum_{n=1}^{N} I_n f_n(s) \qquad (2.25)$$

where s is a parameter for the scatterer cross-section C and f_n is an expansion function. The I_n are unknown weights for the approximate current solution. The span of the expansion functions f_n is the trial subspace in which the numerical solution lies. The expansion functions are generally defined on a discrete geometrical representation, or mesh, for the scatterer surface. The average length of the elements of the mesh is the discretization length h. The dimensionless discretization density or elements per wavelength is

$$n_\lambda = \frac{\lambda}{h} \qquad (2.26)$$

where $\lambda = 2\pi/k_0$.

Substituting (2.25) into (2.8) and testing the integral equation with another set of functions $t_n(s)$ produces the linear system

$$\sum_n Z_{mn} I_n = V_m^i \qquad (2.27)$$

where

$$Z_{mn} = \frac{1}{h} \int ds\, t_m(s) \mathcal{L} f_n \qquad (2.28)$$

$$V_n^i = h^{-1} \int ds\, t_m(s) E^{\text{inc}}(s) \qquad (2.29)$$

where $E^{\text{inc}}(s)$ is a notational simplification for $E^{\text{inc}}[\rho(s)]$. By solving (2.27), the unknown coefficients in the approximate solution $\hat{J}$ for the current on the scatterer can be determined. This can be done by a direct solution of the linear system, typically by LU factorization [10] of the moment matrix $\mathbf{Z}$, or by making use of an iterative linear system solution algorithm. Once an approximation to the current on the scatterer has been obtained, derived quantities such as impedances or far fields can be computed in postprocessing.

The expansion and testing functions are known collectively as basis functions. A basis is a finite set of expansion functions spanning a subspace that approximates the original infinite dimensional space. Basis functions can be nonzero over the entire scatterer surface (entire-domain basis), or each basis function can have support over a small subregion of the scatterer (local basis). Entire-domain basis functions can be chosen to match the oscillatory nature of surface currents in order to reduce the total number of basis functions required to model the current accurately. The difficulty in this case is in the large computational cost required to evaluate the moment matrix integrals in (2.28), since each integration is over the full scatterer surface. Because of this, in practice local basis functions are typically used.

Common types of local basis functions for 2D problems are the delta function, piecewise constant (pulse), and piecewise linear (triangle):

$$f(x) = \delta(x) \tag{2.30a}$$

$$f(x) = \begin{cases} 1 & -h/2 \leq x \leq h/2 \\ 0 & |x| > h/2 \end{cases} \tag{2.30b}$$

$$f(x) = \begin{cases} 1 - |x|/h & -h \leq x \leq h \\ 0 & |x| > h \end{cases} \tag{2.30c}$$

If the mesh is regular, so that each element is of length h, then the basis functions are of the form $f_n(s) = f(s - s_n)$, where s_n is the nth mesh element center point.

This discretization procedure is called the method of moments (MoM), method of weighted residuals, or the Galerkin-Petrov method [11]. If the trial and testing subspaces are identical, then MoM reduces to Galerkin's method. By analogy with the finite element method for differential equations, MoM applied to surface integral equations (also known as boundary integral equations) is also referred to as the boundary element method (BEM), particularly if the discretized operator equation is derived from a bilinear form using the Rayleigh-Ritz procedure.

Throughout this book, the hat on a quantity such as $\hat{J}$ denotes an approximate value obtained from the method of moments. The hat is also used for unit vectors, but the meaning of the notation should be clear from context.

2.4.1 Vector Basis Functions

For 3D radiation and scattering problems involving PEC objects, the unknown quantity is a surface current on a two-dimensional manifold. In order to apply the moment method, vector basis functions are required. The surface current approximation is

$$\mathbf{J}_s(\mathbf{r}) \simeq \hat{\mathbf{J}}_s(\mathbf{r}) = \sum_{n=1}^{N} I_n \mathbf{f}_n(\mathbf{r}) \tag{2.31}$$

where $\mathbf{f}_1, \mathbf{f}_2, \ldots, \mathbf{f}_N$ are vector expansion functions. Vector testing functions are also required, which may be taken to be the same as the expansion functions (Galerkin's method), or a different set of functions can be chosen. In the latter case, point matching or line integrations can be used, which implies that the testing functions include delta functions.

For electromagnetic surface integral equations, the most common vector basis is the Rao-Wilton-Glisson (RWG) function defined on a triangular mesh [12]. Other vector basis functions are described in Section 8.4. On a triangular mesh, the RWG vector basis functions are defined by

$$\mathbf{f}_n(\mathbf{r}) = \begin{cases} \frac{l_n}{2A^+}\boldsymbol{\rho}_n^+ & \text{if } \mathbf{r} \text{ is on the } + \text{ triangle} \\ -\frac{l_n}{2A^-}\boldsymbol{\rho}_n^- & \text{if } \mathbf{r} \text{ is on the } - \text{ triangle} \end{cases} \qquad (2.32)$$

where $+$ and $-$ denote the two triangles that share the nth edge, l_n is the length of the edge, and A^+ and A^- are the areas of the two triangles. The vector $\boldsymbol{\rho}_n^+$ points from the vertex of the $+$ triangle that is not on the edge to $\mathbf{r}$, so that

$$\boldsymbol{\rho}_n^+ = \mathbf{r} - \mathbf{r}_i^+ \qquad (2.33)$$

where $\mathbf{r}_i^+$ is the position vector of the opposite vertex. Similarly, $\boldsymbol{\rho}_n^-$ is the vector from the vertex of the $-$ triangle that is not on the edge to $\mathbf{r}$. The divergence of the RWG function is

$$\nabla \cdot \mathbf{f}_n(\mathbf{r}) = \begin{cases} \frac{l_n}{A^+} & \text{if } \mathbf{r} \text{ is on the } + \text{ triangle} \\ -\frac{l_n}{A^-} & \text{if } \mathbf{r} \text{ is on the } - \text{ triangle} \end{cases} \qquad (2.34)$$

Finite divergence represents an important property of the RWG basis. We have defined the vector functions for a flat mesh element, but the RWG basis can be modified for use on curved patches.

2.5 NUMBER OF UNKNOWNS

The goal of error analysis is to characterize the goodness of the approximate current solution (2.25) and derived quantities such as scattering amplitudes or scattering cross-sections in terms of the choice of basis functions, the number of unknowns or degrees of freedom N, the geometry of the scatterer, and other aspects of the numerical method.

The simplest consideration is that the incident field on the scatterer is oscillatory with period on the order of the electromagnetic wavelength λ, and this induces a similar oscillatory behavior for the surface current. According to the Nyquist sampling theorem, at least two sample points per wavelength are required to model a bandlimited function. As a consequence, the number of unknowns required to model an oscillatory surface current is proportional to the electrical size of the scatterer. The total number of unknowns is therefore on the order of

$$N = c_1(k_0 d)^{\dim-1} \qquad (2.35)$$

where dim is the dimensionality of the scattering problem and d is a linear measure of the scatterer. Sharp corners or other singular geometrical features of the scatterer introduce variations in the surface current with higher spatial frequencies, so in most cases some degree of oversampling relative to the electromagnetic wavelength is required. A common rule of thumb is ten unknowns per wavelength.

2.6 SCATTERING AMPLITUDE, SCATTERING WIDTH, AND RADAR CROSS-SECTION

While the surface current is the direct result of a method of moments computation, derived quantities such as scattered fields or scattering amplitudes are often the final desired result of a numerical simulation. Let the incident electric field on a scatterer be a plane wave of the form

$$\mathbf{E}^{\text{inc}}(\mathbf{r}) = \mathbf{E}_0 e^{-j\mathbf{k}^{\text{inc}}\cdot\mathbf{r}} \tag{2.36}$$

where the constant vector $\mathbf{E}_0$ gives the polarization and magnitude of the field, and $\mathbf{k}^{\text{inc}}$ is the wavevector. The scattering amplitude S is then defined by [13]

$$E^{\text{sca}}(\hat{k}^{\text{sca}}, \hat{k}^{\text{inc}}) \to \frac{e^{-jkr}}{kr} S(\hat{k}^{\text{sca}}, \hat{k}^{\text{inc}}) E^{\text{inc}}, \quad r \to \infty \tag{2.37}$$

This complex quantity essentially specifies the amplitude of the scattered field in the direction $\hat{k}^{\text{sca}}$. The scattering cross-section or radar cross-section (RCS) is

$$\sigma(\hat{k}^{\text{sca}}, \hat{k}^{\text{inc}}) = \frac{4\pi}{k_0^2}|S|^2 \tag{2.38}$$

For 2D problems, the scattering amplitude satisfies the definition

$$E^{\text{sca}}(\phi^{\text{sca}}, \phi^{\text{inc}}) \to \sqrt{\frac{2j}{\pi k_0 \rho}} e^{-jk_0\rho} S(\phi^{\text{sca}}, \phi^{\text{inc}}), \quad \rho \to \infty \tag{2.39}$$

where ϕ^{sca} and ϕ^{inc} are the angles of the propagation directions of the incident and scattered fields.

Since surface integral equations model equivalent currents on a scatterer, it is convenient to express the scattering amplitude in terms of currents. For 2D problems with TM polarized fields, the scattered field is given by the radiation integral

$$E_z^{\text{sca}}(\boldsymbol{\rho}) = -\frac{k_0\eta}{4} \int_C ds' \, H_0^{(2)}(k_0|\boldsymbol{\rho} - \boldsymbol{\rho}'|) J_z(\boldsymbol{\rho}') \tag{2.40}$$

Using the large argument expansion of the Hankel function and the far field approximation $|\boldsymbol{\rho} - \boldsymbol{\rho}'| \simeq \rho - \hat{\rho} \cdot \boldsymbol{\rho}'$, we obtain the far field radiation integral

$$E_z^{\text{sca}}(\boldsymbol{\rho}) = -\frac{k_0\eta}{4} \sqrt{\frac{2j}{\pi k_0 \rho}} e^{-jk_0\rho} \int_C ds' \, e^{jk_0\hat{\rho}\cdot\boldsymbol{\rho}'} J_z(\boldsymbol{\rho}'), \quad \rho \to \infty \tag{2.41}$$

This expression can be interpreted as a Fourier integral. The scattered field in the $\hat{\rho}$ direction depends on the Fourier transform of $J_z(\boldsymbol{\rho}')$ evaluated at $k_0\hat{\rho}$. In other words, the integral above can be written as

$$\int_C ds'\, e^{jk_0\hat{\rho}\cdot\boldsymbol{\rho}'} J_z(\boldsymbol{\rho}') = \tilde{J}_z(k_0\hat{\rho}) \tag{2.42}$$

where $\tilde{J}_z(\mathbf{k})$ is the Fourier transform of $J_z(\boldsymbol{\rho})$.

As a function of $\boldsymbol{\rho}'$, the complex conjugate of the exponential in the integrand has the form of a plane wave propagating in the $\hat{\rho}$ direction. We introduce the notation

$$E^s(\boldsymbol{\rho}) = e^{-jk_0\hat{\rho}\cdot\boldsymbol{\rho}} \tag{2.43}$$

for this plane wave, so that E_z^{sca} is the scattered electric field and E_z^s is a plane wave propagating away from the scatterer. This leads to

$$E_z^{\text{sca}}(\boldsymbol{\rho}) = -\frac{k_0\eta}{4}\sqrt{\frac{2j}{\pi k_0\rho}}\, e^{-jk_0\rho} \int_C ds'\, E^{s*}(\boldsymbol{\rho}') J_z(\boldsymbol{\rho}'), \qquad \rho \to \infty$$

The integral can be identified as the L^2 inner product $\langle E^s, J_z\rangle$ on the scatterer C. This inner product by definition includes a complex conjugate on the first of the two functions in the inner product. It is important to be aware that while this notation is standard in functional analysis, in the electromagnetics literature the angle bracket is commonly defined without the complex conjugate to indicate the symmetric product or reaction.

Using the L^2 inner product, we have

$$E_z^{\text{sca}}(\boldsymbol{\rho}) = -\frac{k_0\eta}{4}\sqrt{\frac{2j}{\pi k_0\rho}}\, e^{-jk_0\rho} \langle E^s, J_z\rangle, \qquad \rho \to \infty$$

Comparing this expression with (2.39) shows that the scattering amplitude is given by

$$S(\phi^{\text{inc}}, \phi^{\text{sca}}) = -\frac{k_0\eta}{4}\langle E^s, J_z\rangle \tag{2.44}$$

Similar results can be obtained for the TE polarization and for 3D problems.

In the method of moments framework, the bistatic scattering amplitude can be obtained from the approximate current $\hat{J}$ by substituting (2.25) into (2.44) to obtain

$$\hat{S}(\phi^{\text{inc}}, \phi^{\text{sca}}) = -\frac{k_0\eta}{4}\left\langle E^s, \sum_{n=1}^{N} I_n f_n\right\rangle$$

$$= -\frac{k_0\eta}{4}\sum_{n=1}^{N} V_n^{s*} I_n \tag{2.45}$$

where

$$V_n^s = \int_C ds\, E^s(s) f_n(s) \tag{2.46}$$

From this it can be seen that the plane wave E_s is discretized on the surface of the conductor using the expansion functions f_n in a manner similar to (2.29) for the incident field. In matrix notation, the scattering amplitude can be written as

$$\hat{S}(\phi^{\mathrm{inc}}, \phi^{\mathrm{sca}}) = -\frac{k_0 \eta}{4} \mathbf{v}^{sH} \mathbf{Z}^{-1} \mathbf{v}^i \tag{2.47}$$

The simple symmetry of this expression is another manifestation of the special relationship between the scattering amplitude and the method of moments.

2.7 ERROR MEASURES

The direct result of a MoM computation is an approximation $\hat{J}$ to the surface current on the scatterer. Of course, the goal is to obtain an approximation that is as close as possible to the exact current solution. In applications, other quantities are derived from the current in postprocessing, including near fields, far fields, scattering amplitudes, radar cross-section (RCS), and port impedances. Here, we will list some of the common error measures for surface currents, scattering amplitudes, and RCS.

For smooth scatterers, if no sources lie on the scatterer surface, then the surface current is bounded and hence is square integrable. In this case, the L^2 error of the numerical current solution $\hat{J}$ with respect to the exact current J is in principle a meaningful error measure, which could be used to assess the accuracy of a numerical method. The L^2 current solution error is

$$\mathrm{Err}_{L^2} = \left\| \hat{J} - J \right\| = \left[\int_S ds \left| \hat{J}(\mathbf{r}) - J(\mathbf{r}) \right|^2 \right]^{1/2} \tag{2.48}$$

The relative L^2 error is

$$\mathrm{Err}_{L^2,\mathrm{rel}} = \frac{\left\| \hat{J} - J \right\|}{\left\| J \right\|}$$

$$= \left[\int_S ds \left| \hat{J}(\mathbf{r}) - J(\mathbf{r}) \right|^2 \right]^{1/2} \left[\int_S ds \left| J(\mathbf{r}) \right|^2 \right]^{-1/2} \tag{2.49}$$

which has the advantage of being dimensionless.

We can also measure the current error at a finite number of sample points using the discrete root mean square (RMS) current error

$$\mathrm{Err}_{\mathrm{RMS}} = \left[\frac{1}{N} \sum_{n=1}^{N} \left| \hat{J}_n - J_n \right|^2 \right]^{1/2} \tag{2.50}$$

The relative RMS error is

$$\text{Err}_{\text{RMS,rel}} = \left[\sum_{n=1}^{N} |\hat{\mathbf{J}}_n - \mathbf{J}_n|^2\right]^{1/2} \left[\sum_{n=1}^{N} |\mathbf{J}_n|^2\right]^{-1/2} \tag{2.51}$$

where $\mathbf{J}_n$ is the surface current density vector evaluated at the nth node point of the surface mesh. The subscript rel will be dropped in later chapters for notational simplicity. A minor benefit of discrete error measures is that samples are often readily available as part of a moment method computation, whereas the continuous function must be obtained through (2.25). Except in a few places where the L^2 norm is more convenient, for reasons to be examined shortly we will use relative RMS error throughout the book.

One drawback of measuring moment method solution error using the surface current is that if the scatterer has an edge or corner, then the current can be singular, and the L^2 norm may not exist. Since the RMS value is a sampled or discrete L^2 norm, computing the RMS value of currents for problems with singularities is not appropriate, at least for convergence studies where the numerical current approaches the exact singularity as the mesh is refined and the RMS value may not converge. For scatterers with singularities, the current error for singular regions of the scatterer must be measured using a different norm such as L^1 or a Sobolev norm, as discussed in Section 1.3.

Another approach is to consider the error in a scattering amplitude or radar cross-section. The relative scattering amplitude error for given incident and scattered directions is

$$\text{Err}_S = \frac{|\hat{S} - S|}{|S|} \tag{2.52}$$

where $\hat{S}$ is the computed scattering amplitude obtained from the approximate current solution. Error measures that take into account more than one scattered direction are also commonly used in computational electromagnetics. These include the maximum relative RCS error

$$\text{Err}_{\text{max}} = \max_{1 \leq m \leq M} |\hat{\sigma}(\theta_m, \phi_m)/\sigma(\theta_m, \phi_m) - 1| \tag{2.53}$$

and the relative RMS dB RCS error

$$\text{Err}_{\text{RMS}} = \left\{\frac{1}{M} \sum_{m=1}^{M} |10 \log_{10}[\hat{\sigma}(\theta_m, \phi_m)/\sigma(\theta_m, \phi_m)]|^2\right\}^{1/2} \tag{2.54}$$

where the angles θ_m and ϕ_m represent M selected scattering directions. In some cases, the scattering amplitude or RCS errors may be small even when the surface current error is large. This effect will be analyzed in detail in later chapters.

2.8 Basic Concepts of Modal Error Analysis

As a warmup to the error analysis of the full method of moments, we will consider here the error incurred when expanding a Fourier mode using a set of basis functions. Let $u(x)$ be a function defined on the interval $[0, d]$. We wish to determine the error in an expansion of the form (2.25), so that

$$\hat{u}(x) = \sum_{n=1}^{N} a_n f_n(x) \qquad (2.55)$$

With the method of moments, the expansion coefficients a_n are determined by solving a linear system. Here, we will simplify the problem by taking $u(x)$ to be a Fourier mode of the form

$$u(x) = e^{-jkx} \qquad (2.56)$$

and determining the coefficients in (2.55) by either interpolation or projection. The expansion functions will be pulse functions as given by (2.30b) and the mesh will consist of N regularly spaced elements with center points $x_n = (n - 1/2)h$, where $h = d/N$.

2.8.1 Interpolation Error

We will first take a_n to be samples of u at the element centers, so that $a_n = u(x_n)$. Because the pulse functions $f(x - x_n)$ are equal to one at x_n, $\hat{u}(x)$ is equal to $u(x)$ at the element center points. The RMS error (2.50) is therefore zero when measured at the element centers.

As shown in Figure 2.1, because of interpolation error caused by the basis functions, the L^2 norm of the difference between the exact and interpolated modes is nonzero. The L^2 error is

$$\|\hat{u} - u\|^2 = \int \left| \sum_n a_n f_n(x) - u(x) \right|^2 dx$$
$$= 2d \left[1 - \frac{1}{h} \int e^{jkx} f(x)\, dx \right]$$
$$= 2d[1 - F(k)] \qquad (2.57)$$

where

$$F(k) = \frac{1}{h} \int e^{jkx} f(x)\, dx$$
$$= \frac{\sin(kh/2)}{kh/2} \qquad (2.58)$$

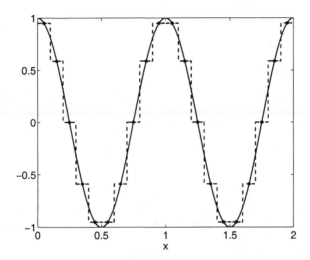

Figure 2.1: A Fourier mode and an expansion using pulse basis functions. The spatial frequency is $k = 2\pi$ and the number of mesh elements is $N = 20$. The RMS error at the element centers is zero, since the expanded function and the exact function are equal, but the L^2 error is nonzero due to the stairstepping effect caused by the pulse functions.

This is the Fourier transform of the canonical expansion function centered at $x = 0$, scaled such that $F(0) = 1$. The relative L^2 error is

$$\frac{\|\hat{u} - u\|}{\|u\|} = |2[1 - F(k)]|^{1/2} \tag{2.59}$$

since the squared L^2 norm of (2.56) on $[0, d]$ is d.

If we expand $F(k)$ for small h, we obtain for the squared L^2 error

$$\|\hat{u} - u\|^2 \simeq \frac{k^2 h^2}{12} d \tag{2.60}$$

The relative squared L^2 error can be approximated by

$$\frac{\|\hat{u} - u\|^2}{\|u\|^2} \simeq \frac{k^2 h^2}{12} \tag{2.61}$$

and the relative L^2 error becomes

$$\frac{\|\hat{u} - u\|}{\|u\|} \simeq \frac{kh}{\sqrt{12}} \tag{2.62}$$

As the number of elements is increased, the L^2 error is first order in the mesh element size h.

2.8.2 Mesh Nyquist Frequency

If the spatial frequency k is increased, the expansion (2.60) eventually breaks down. Returning to (2.58), we can see that the relative L^2 error approaches 100% when the sinc function in (2.58) is significantly different from one. This occurs for spatial frequencies larger than

$$k_{max} = \frac{\pi}{h} \qquad (2.63)$$

This is a measure of the largest spatial frequency that can be represented on the mesh. The associated wavelength is $\lambda_{min} = h/2$, from which we can see that the real and imaginary parts of the mode alternate in sign from one element to the next. This is the maximum spatial frequency that can be represented using discrete, evenly spaced samples, so we will refer to k_{max} as the mesh Nyquist frequency.

2.8.3 Projection Error

When an integral operator or incident field is discretized according to (2.28) and (2.29), the operator kernel or field is projected onto the basis functions by integration. If we obtain the coefficients a_n in (2.55) by this type of projection, then

$$a_n = \frac{1}{h} \int u(x) f_n(x) \, dx \qquad (2.64)$$

Using (2.56) and (2.58), this evaluates to

$$a_n = u(x_n) F(-k) \qquad (2.65)$$

where $F(k) = F(-k)$ for symmetric basis functions. The samples are equal to the exact mode evaluated at the node points, but with an additional scale factor given by the Fourier transform of the expansion function at the spatial frequency of the mode. We will refer to this multiplicative factor as projection error. If the spatial frequency of the mode is small, then $F(k) \simeq 1$, and the scaling caused by projection error is small. If the spatial frequency is larger than k_{max}, then the projection error is large, because the sinc function in (2.58) decreases with k and is less than unity in magnitude.

If we compute the L^2 error of $\hat{u}$ with the expansion coefficients (2.65), the error measure includes contributions from projection error as well as the interpolation error analyzed in the previous section. This can be seen in Figure 2.2. The relative L^2 error in this case is

$$\frac{\|\hat{u} - u\|}{\|u\|} = |1 - F(-k)F(k)|^{1/2} \qquad (2.66)$$

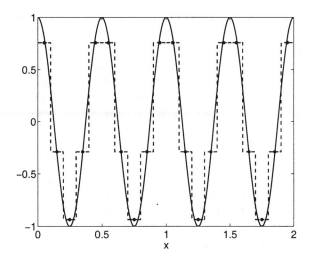

Figure 2.2: A Fourier mode and an expansion using pulse basis functions, where the mode is projected onto the basis functions to obtain the expansion coefficients. The spatial frequency is $k = 4\pi$ and the number of mesh elements is $N = 20$. The RMS error at the element centers is nonzero, because the expansion coefficients are slightly different from the exact value of the mode at the element centers. The L^2 error is of the same order as the error that would be obtained with exact interpolation of the mode, but the RMS error is smaller.

By expanding this expression for small h, it can be seen that the L^2 error for the projected mode has the same order in h as (2.59) for the L^2 error with exact interpolation and no projection error. Since the order of the L^2 error is the same whether the approximation $\hat{u}$ is obtained by exact interpolation of the mode at the node points or by projection of the mode onto the expansion functions, we conclude that L^2 error conflates interpolation and projection errors.

The discrete RMS error (2.50) at the mesh element centers, on the other hand, is more straightforward to understand because it isolates the projection error. The relative RMS error is simply

$$\text{Err}_{\text{RMS}} = |F(k) - 1| \tag{2.67}$$

because the scale factor is identical for each sample value. When expanded for small h, the relative RMS error becomes

$$\text{Err}_{\text{RMS}} \simeq \frac{k^2 h^2}{24} \tag{2.68}$$

which is similar to (2.61) for the squared L^2 error. The projection error as measured with the discrete RMS value is second order in h.

In view of these results, it is apparent that for a given implementation of the method of moments the L^2 and discrete RMS error measures are in general different. This explains some of the disagreement in the literature on moment method solution convergence rates. Because the RMS error at interpolatory sample points isolates projection error from interpolation error, we will use RMS error almost exclusively throughout this book to measure current solution error. The exception will be in Section 8.3 when considering higher-order orthogonal (and noninterpolatory) polynomial basis functions.

Since the eigenfunctions of the integral operators of electromagnetics for many scattering problems are Fourier modes of the form of (2.56) or can be approximated as such, the analyses of the following chapters will make heavy use of the simple concepts considered in this section, particularly the mesh Nyquist frequency and projection error. In fact, the second-order error in (2.68) is essentially the RMS current solution error that will be obtained in the following chapter for the method of moments applied to a smooth scatterer with pulse expansion functions.

REFERENCES

[1]   W. C. Chew, *Waves and Fields in Inhomogeneous Media*. New York: IEEE Press, 1995.

[2]   G. C. Hsiao and R. E. Kleinman, "Mathematical foundations for error estimation in numerical solutions of integral equation in electromagnetics," *IEEE Trans. Ant. Propag.*, vol. 45, pp. 316–328, Mar. 1997.

[3]   C. A. Balanis, *Advanced Engineering Electromagnetics*. New York: John Wiley & Sons, 1989.

[4]   D. R. Wilton, "Review of current status and trends in the use of integral equations in computational electromagnetics," *Electromagnetics*, vol. 12, pp. 287–341, 1992.

[5]   W. C. Chew and J. M. Song, "Gedanken experiments to understand the internal resonance problems of electromagnetic scattering," *Electromagnetics*, vol. 27, pp. 457–471, Nov. 2007.

[6]   C. P. Davis and K. F. Warnick, "Error analysis of 2D MoM for MFIE/EFIE/CFIE based on the circular cylinder," *IEEE Trans. Ant. Propag.*, vol. 53, pp. 321–331, Jan. 2005.

[7]   A. Helaly and H. M. Fahmy, "Combined-field integral equation," *Electronics Letters*, vol. 29, pp. 1678–1679, September 1993.

[8]   F. K. Oshiro, K. M. Mitzner, S. S. Locus, and et al., "Calculation of radar cross section," Tech. Rep. AFAL-Tr-70-21, part II, Air Force Avionics Laboratory, April 1970.

[9]   R. Kress, "Minimizing the condition number of boundary integral operators in acoustic and electromagnetic scattering," *Q. Jl. Mech. Appl. Math.*, vol. 38(2), pp. 323–341, 1985.

[10] G. Strang, *Introduction to Linear Algebra*. Wellesley-Cambridge Press, 1993.

[11] A. G. Dallas, G. C. Hsiao, and R. E. Kleinman, "Observations on the numerical stability of the Galerkin method," *Adv. Comput. Math.*, vol. 9, pp. 37–67, 1998.

[12] S. M. Rao, D. R. Wilton, and A. W. Glisson, "Electromagnetic scattering by surfaces of arbitrary shape," *IEEE Trans. Ant. Propag.*, vol. 30, pp. 409–418, May 1982.

[13] J. J. Bowman, T. B. A. Senior, and P. L. E. Uslenghi, *Electromagnetic and Acoustic Scattering by Simple Shapes*. New York: Hemisphere, 1987.

Chapter 3

Error Analysis of the EFIE

with Weng Cho Chew

In this chapter, we will develop many of the ideas that will be used throughout the book in the context of a simple 2D scattering problem. The most common test cases used to validate numerical simulations are the circular PEC cylinder and sphere. Normally, a simulation code is used to generate scattered fields or surface current for the test geometry, and the error in the numerical solution relative to an exact, analytical solution is computed. What we will do here is somewhat different: we will apply the method of moments algorithm for the EFIE to the circular cylinder analytically, rather than numerically. In other words, we will derive a series expression not for the exact solution for the cylinder, but for the approximate numerical solution itself. Armed with such an analytical expression for the approximate solution, we can use mathematical insight to understand the impact of different aspects of the numerical method on the solution error.

Even though the circular cylinder is a very simple scattering problem, careful analysis of this case will provide error estimates for surface currents and scattering amplitudes, including contributions from quadrature error associated with numerical integration of moment matrix elements. Since the circular cylinder is closed, we can also gain significant insight into the behavior of numerical solutions near internal resonance frequencies. In fact, the only major physical effect that we will not be able to study with this geometry is the effect of geometric singularities such as edges and corners, since the circular cylinder is a smooth scatterer.

In this chapter, we will first consider an ideal implementation of the method of moments, for which:

- Moment matrix element integrations are evaluated exactly.

- A conformal and regular (identical mesh elements) geometrical representation of the scatterer is employed.

- The linear system produced by the method of moments is inverted exactly.

With an ideal implementation, under normal circumstances the minimum possible solution error for a given set of basis functions is achieved. In some cases, a nonideal implementation can actually be more accurate than an ideal discretization, but generally the ideal implementation can be viewed as setting a lower bound on error.

After analyzing the ideal case, we will first relax the assumption of exact integration of moment matrix elements, and introduce quadrature error into the analysis. At this point, the concept of variationality of the method of moments will enter into the discussion. The curved mesh will then be replaced by a flat-facet mesh, in order to assess the impact of geometrical discretization error. Consideration of the third aspect of a nonideal implementation, linear system solution error, will be deferred to Chapter 9.

3.1 TM-EFIE with Ideal Discretizations

We will begin with a spectral decomposition in terms of eigenvalues and eigenfunctions for the EFIE operator. In general, such a decomposition is available if and only if the operator is normal (a matrix is normal if it commutes with its Hermitian conjugate and an operator is normal if it commutes with its adjoint [1]). For general scatterer geometries, the EFIE operator is nonnormal, so a spectral decomposition is not available, although a singular value decomposition does exist [2]. In later chapters we will sidestep this problem by proving that the EFIE operator can be approximated by a normal operator. For the special case of the circular cylinder, however, the EFIE operator is normal and an exact spectral decomposition can be employed in the error analysis. The starting point for the spectral decomposition is a cylindrical mode expansion of the 2D free space Green's function,

$$H_0^{(2)}(k_0|\boldsymbol{\rho} - \boldsymbol{\rho}'|) = \sum_{q=-\infty}^{\infty} J_q(k_0 a) H_q^{(2)}(k_0 a) e^{-jq(\phi-\phi')} \tag{3.1}$$

where $\boldsymbol{\rho}$ and $\boldsymbol{\rho}'$ lie on a circle of radius a and ϕ and ϕ' are the corresponding angles in the cylindrical coordinate system. The above is a Fourier series expansion of the Green's function on the circle $\rho = \rho' = a$ in the $\phi - \phi'$ variable. By substituting this expression into (2.9), a spectral decomposition of the integral operator is immediately obtained.

From (3.1), it can be seen that the eigenfunctions of the integral operator $\mathcal{L}$ are the Fourier modes $e^{-jq\phi}$, so the action of the EFIE operator on an arbitrary function is to scale each Fourier component by the eigenvalue

$$\lambda_q = \tfrac{1}{2}\eta\pi k_0 a J_q(k_0 a) H_q^{(2)}(k_0 a) \tag{3.2}$$

where the leading constant is the circumference of the scatterer, $2\pi a$, multiplied by the leading constant in (2.9). We will pause here to consider the physical meaning of some of the mathematical properties of the operator spectrum.

Normality. From an operator theory point of view, the existence of a spectral decomposition of this form indicates that $\mathcal{L}$ is a normal operator, although $\mathcal{L}$ is not self-adjoint, because the eigenvalues are complex. Normality of a system in general is associated with the existence of a set of orthogonal modes, each of which can be excited independently. For other scatterer geometries, it can be shown numerically that $\mathcal{L}$ is nonnormal. We will discuss the physical meaning of nonnormality later in the context of the flat strip in Chapter 5.

Smoothing property of the TM-EFIE operator. Using the large-order expansion

$$J_\nu(x)H_\nu^{(2)}(x) \sim \frac{j}{\pi|\nu|}, \quad \nu \to \infty \tag{3.3}$$

the eigenvalues of the integral operator for large orders decay in magnitude as

$$\lambda_q \sim \frac{1}{|q|} \tag{3.4}$$

The eigenvalues therefore accumulate in the complex plane at the origin. This is a manifestation of the fact that the operator $\mathcal{L}$ is compact, which loosely speaking means that the operator maps a surface current J_s with bounded L_2 norm to a tangential field $\mathcal{L}J_s$ with bounded L_2 norm. For highly oscillatory components of the surface current, due to (3.4), the amplitude of those components is reduced by the integral operator, so that the resulting tangential field is smoother than the current. For this reason, $\mathcal{L}$ can be considered an integrating or smoothing operator.

While the operator $\mathcal{L}$ does have a smoothing property, due to the singular kernel the smoothing effect is weaker than for other first-kind integral operators with nonsingular kernels. An operator defined by convolution with a bounded, square-integrable kernel function $k(x)$ according to

$$\mathcal{K}u = \int_a^b k(x - x')u(x')\,dx' \tag{3.5}$$

is necessarily a smoothing operator, which can be seen by taking the Fourier transform of the above equation. If the kernel $k(x)$ is nonsingular, the Fourier series coefficients of the kernel decay more rapidly than (3.4) and the operator reduces the amplitude of rapidly varying Fourier components of u. Because of the logarithmic singularity of the kernel of the TM-EFIE, the eigenvalues fall off in magnitude relatively slowly with order, leading to a weaker smoothing property than (3.5). We will see in Chapter 9 that this has important ramifications for the conditioning of the method of moments linear system.

Radiating and nonradiating modes.   The spatial frequency of the qth mode is q/a. The transition to the asymptotic behavior specified by (3.4) is $|q| \simeq k_0 a$. Physically, larger orders correspond to highly oscillatory current modes that do not radiate significant amounts of power into the far field, or nonradiating modes. Lower orders correspond to currents that radiate strongly into the far field, or radiating modes. In the complex plane, the eigenvalues of the radiating modes are distributed in the right half plane, whereas the eigenvalues of the nonradiating modes lie near the imaginary axis on an arc extending in the limit to the origin.

Radiating and nonradiating modes can be understood by analogy with an aperture antenna. If the field distribution across an aperture is uniform, then the radiation pattern of the aperture has a main lobe in the broadside direction, or the direction normal to the aperture plane. In order to steer the beam, the aperture field distribution can be given a phase progression of the form e^{-jkx}. The steering angle of the beam is determined by the phase coefficient k. For $k = 0$, the aperture distribution is uniform. As k increases, the main lobe direction is steered away from broadside. At $k = k_0$, the main lobe is steered to the endfire direction, or parallel to the aperture plane. For $|k| > k_0$, the aperture distribution becomes nonradiating, in the sense that the amount of real power radiated to the far field decreases exponentially as k increases. While nonradiating modes radiate very weak far fields, they do produce near fields that store reactive energy.

3.1.1 Discretized Operator Spectrum

For the purpose of error analysis, we are interested primarily in the discretized matrix operator used to obtain an approximate numerical solution. Our immediate goal is to determine the eigenvalues of the discretized operator in terms of the continuous operator eigenvalues. The eigenvalues of the discretized operator will then be used to determine error in numerical solutions for the current and scattering amplitude.

To accomplish this, we will insert (3.1) into (2.28) for the elements of the moment matrix. To simplify the analysis, we will assume that the expansion and testing functions used as basis sets for the method of moments procedure are shift invariant, so that $f_n(\phi) = f(\phi - \phi_n)$ and $t_n(\phi) = t(\phi - \phi_n)$, where $f(\phi)$ and $t(\phi)$ are canonical basis functions located at $\phi = 0$. The nodes are taken to be evenly spaced points around the circumference of the scatterer, so that $\phi_n = (n-1/2)\theta_0$, where $\theta_0 = 2\pi/N$ and N is the total number of basis functions. It is worth noting that the mesh element length in this case is $h = 2\pi a/N$, and the dimensionless mesh density is $n_\lambda = \lambda/h = N/(k_0 a)$.

Under these assumptions, the moment matrix becomes

$$Z_{mn} = \frac{\eta \pi k_0 a}{2N} \sum_{q=-\infty}^{\infty} J_q(k_0 a) H_q^{(2)}(k_0 a) T_{-q} F_q e^{-jq(\phi_m - \phi_n)} \tag{3.6}$$

where T_q represents the Fourier series coefficients of $t(\phi)$,

$$T_q = \frac{1}{\theta_0} \int d\phi \, t(\phi) e^{jq\phi} \tag{3.7}$$

evaluated at q and normalized by $1/\theta_0$. F_q for the expansion basis function is defined similarly by

$$F_q = \frac{1}{\theta_0} \int d\phi \, f(\phi) e^{jq\phi} \tag{3.8}$$

This is the Fourier transform in (2.58) evaluated at the discrete set of spatial frequencies given by $k_q = q/a$.

Equation (3.6) indicates that $Z_{mn} = Z_{m-n}$, and hence Z_{mn} is a Toeplitz matrix. By inspection of (3.6), the eigenvectors of the moment matrix are of the form $e^{-jq\phi_n}$, where q is an integer and n indexes the components of the eigenvector. The corresponding eigenvalues can be determined by forming a matrix-vector multiplication of the moment matrix in the form (3.6) and the eigenvector:

$$\sum_{n=1}^{N} Z_{mn} e^{-jq\phi_n} = \frac{\eta \pi k_0 a}{2N} \sum_{r=-\infty}^{\infty} J_r(k_0 a) H_r^{(2)}(k_0 a) T_{-r} F_r e^{-jr\phi_m} \sum_{n=1}^{N} e^{-j(q-r)\phi_n} \tag{3.9}$$

The sum over n can be evaluated using the identity

$$\sum_{n=1}^{N} e^{jr\phi_n} = (-1)^r \frac{\sin(\pi r)}{\sin(\pi r/N)} \tag{3.10}$$

The right-hand side of (3.10) is equal to $(-1)^s N$ if $r = sN$, where s is an integer, and vanishes otherwise. Equation (3.9) then becomes

$$\sum_{n=1}^{N} Z_{mn} e^{-jq\phi_n} = \left[\frac{\eta \pi k_0 a}{2} \sum_{s=-\infty}^{\infty} J_{q+sN}(k_0 a) H_{q+sN}^{(2)}(k_0 a) T_{-q-sN} F_{q+sN} \right] e^{-jq\phi_m} \tag{3.11}$$

where we have assumed that N is even. From this expression, we can identify

$$\hat{\lambda}_q = \frac{\eta \pi k_0 a}{2} \sum_{s=-\infty}^{\infty} J_{q+sN}(k_0 a) H_{q+sN}^{(2)}(k_0 a) T_{-q-sN} F_{q+sN} \tag{3.12}$$

as the eigenvalue corresponding to the eigenvector $e^{-jq\phi_n}$. In terms of the exact eigenvalue (3.2), this can be expressed as

$$\hat{\lambda}_q = \sum_{s=-\infty}^{\infty} \lambda_{q+sN} T_{-q-sN} F_{q+sN} \tag{3.13}$$

The moment matrix, being square with N rows and columns, has N eigenvalues. It is not immediately obvious that this is reflected in (3.13), but under closer inspection, it can be seen that the expression for $\hat{\lambda}_q$ is periodic with period N.

The eigenvalues of the moment matrix for an example case are shown in Figure 3.1. The low-order eigenvalues with small values of the index q are distributed in the right half plane. These are the radiating current modes. For large values of q, the eigenvalues approach the origin along the positive imaginary axis. These are the nonradiating current modes that oscillate more rapidly than the wavenumber k_0. From Poynting's theorem, the real part of the eigenvalue is the real power radiated by the mode, and the imaginary part is proportional to the difference between the stored electric and magnetic energy (see Chapter 6). The nonradiating modes with eigenvalues near the imaginary axis have only a very small real part, as expected.

The spectrum of the continuous operator is similar to that of the moment matrix, except that there are a countably infinite number of eigenvalues that approach the origin as an accumulation point, and there is a small shift or difference in the values of corresponding eigenvalues. This difference is spectral error and will turn out to be crucial in determining the solution error of the method of moments. We will now look more deeply into the relationship between the eigenvalues of the moment matrix and those of the continuous integral operator.

3.1.2 Comparison of the Discretized and Exact Operator Spectra

The most basic fact about the exact and approximate eigenvalues is that the N eigenvalues and eigenvectors of the moment matrix are approximations to N of the eigenvalues and eigenvectors of the continuous integral operator $\mathcal{L}$. One simple manifestation of this is that (3.13) approaches (3.2) as the number of basis functions N becomes large. Under rather weak assumptions about the expansion and testing functions, in the limit as $N \to \infty$, all terms in the sum in (3.13) vanish except the $s = 0$ term, and $T_{-q} F_q \to 1$, from which it follows that $\hat{\lambda}_q \to \lambda_q$ as $N \to \infty$. This accounts for the proximity of the two sets of eigenvalues shown in Figure 3.1.

Furthermore, from the treatment in the previous section, the eigenvectors associated with the moment matrix eigenvalues are merely sampled versions of the continuous eigenfunctions. Since (3.13) is periodic, it suffices to consider the range $|q| \leq N/2$ (we will ignore the unimportant technicalities associated with N being even or odd). Setting $q = 0$ in the eigenvector $e^{-jq\phi_n}$ corresponds to a vector of samples around the

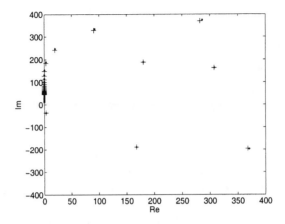

Figure 3.1: Spectrum of the EFIE operator, TM polarized incident field, circular cylinder scatterer with radius $k_0 a = 2\pi$. Pluses: moment matrix spectrum, mesh density $n_\lambda = 10$. Dots: exact eigenvalues of the continuous operator.

cylinder equal to unity, or in other words, a DC current mode. As the magnitude of q increases, the mode becomes oscillatory. The spatial frequency is

$$k_q = \frac{q}{a} \tag{3.14}$$

The eigenvector with the largest spatial frequency will be of particular importance. This corresponds to an order of $q = \pm N/2$, and the mode spatial frequency is identical to the mesh Nyquist frequency $k_{\max}$ given by (2.63).

The domain of the continuous integral operator includes functions with essentially arbitrary spatial frequency content, ranging from slowly varying functions to highly oscillatory functions. The discrete operator represented by the moment matrix, on the other hand, represents only the low frequency portion of the infinite "bandwidth" of the continuous operator. This is closely associated with the approximating capability of the expansion and testing functions. With localized basis functions, highly oscillatory functions with spatial frequency above $k_{\max}$ cannot be approximated, so these functions are not represented in the spectrum of the moment matrix. For the low frequency part of the spectrum (i.e., current modes $e^{-jq\phi_n}$ with $|q| \leq N/2$), we can make a correspondence between the N lowest order eigenvalues of the continuous operator and the N eigenvalues of the moment matrix. We refer to the eigenvalues of $\mathcal{L}$ with order $|q| > N/2$ as unmodeled eigenvalues, since the spatial frequency of the corresponding eigenfunctions is greater than the mesh Nyquist frequency, and they cannot be represented in the trial subspace spanned by the basis set used to discretize the integral operator. Modes with spatial frequencies below the mesh Nyquist frequency ($|q| \le N/2$) might be referred to as modeled modes.

3.1.3 Spectral Error

We have just observed that there is a correspondence between the N eigenvalues of the moment matrix and the N lowest order eigenvalues of the continuous integral operator. From (3.1), it can be seen that the exact continuous operator and moment matrix eigenvalues are not equal. We will refer to the eigenvalue shift between the exact and approximate eigenvalues as spectral error. It will be seen later that the spectral error has a close relationship to the current and scattering amplitude solution errors. Therefore, it will be useful to understand the causes and physical meaning of the spectral error in more detail.

The moment matrix eigenvalue (3.13) can be rearranged into the form

$$\hat{\lambda}_q = \lambda_q T_{-q} F_q + \sum_{s \ne 0} \lambda_{q+sN} T_{-q-sN} F_{q+sN} \tag{3.15}$$

where we have separated the $s = 0$ term in the sum. It can be seen that the moment matrix eigenvalue $\hat{\lambda}_q$, relative to the exact eigenvalue λ_q, is perturbed by two contributions, one that is multiplicative, and another that is additive and is associated with high-order eigenvalues of the integral operator. The multiplicative change in the eigenvalue we will refer to as projection error, by analogy with the discussion in Section 2.8.3, and the additive part we will refer to as aliasing error.

In terms of the spectral error Δ_q, we can write the approximate eigenvalue in the form

$$\hat{\lambda}_q = \lambda_q + \Delta_q \tag{3.16}$$

where Δ_q is the difference between the exact eigenvalue λ_q of the continuous operator and the corresponding eigenvalue $\hat{\lambda}_q$ of the moment matrix. From (3.15), the spectral error is

$$\Delta_q = \lambda_q (T_{-q} F_q - 1) + \sum_{s \ne 0} \lambda_{q+sN} T_{-q-sN} F_{q+sN} \tag{3.17}$$

It is also convenient to consider the relative spectral error

$$E_q = \frac{\Delta_q}{\lambda_q} = \underbrace{T_{-q} F_q - 1}_{\substack{\text{Projection} \\ \text{error}}} + \underbrace{\frac{1}{\lambda_q} \sum_{s \ne 0} \lambda_{q+sN} T_{-q-sN} F_{q+sN}}_{\text{Aliasing error}} \tag{3.18}$$

which can be divided into two terms corresponding to the multiplicative and additive contributions in (3.15).

The objective at this point is to understand this expression for spectral error in more detail, by considering the physical meaning of the two types of error. We will then develop closed form approximations for the spectral error for specific basis functions. Finally, the spectral error will be used to determine the solution error for the surface current and scattering amplitudes.

Projection error. The first term of the relative spectral error, $T_{-q}F_q - 1$, is caused by projection of the "modeled," low spatial frequency ($|q| \leq N/2$) eigenfunctions of the kernel onto the expansion and testing basis subspaces. This is closely related to the projection error analyzed from a simple point of view in Section 2.8.3, except that here the error arises from projection of the operator kernel onto testing and expansion functions. Since the kernel has two coordinates (the field and source points), the single-mode projection error of Section 2.8.3 is repeated for projection onto the testing and expansion functions.

To further understand this error term, it is helpful to use bra-ket notation and to construct a projection operator onto a finite dimensional basis subspace. Let the qth eigenfunction be denoted by $|u_q\rangle$, so that

$$\mathcal{L} = \sum_q \lambda_q |u_q\rangle \langle u_q| \tag{3.19}$$

We define the testing subspace discretization operator to be

$$\mathcal{P} = \sum_{m=1}^{N} |t_m\rangle \langle t_m| \tag{3.20}$$

If the basis functions are not orthogonal, then this operator is not a projection operator, since $\mathcal{P}^2 \neq \mathcal{P}$, but it can be viewed as a projection operator in an approximate sense. Since the operator is not an exact projection, $|\hat{u}\rangle = \mathcal{P}|u\rangle$ does not necessarily minimize the error $\|\hat{u} - u\|$ over all possible linear combinations $\hat{u}$ of the basis functions t_m. The motivation for defining $\mathcal{P}$ according to (3.20) is that applying this type of projection to a continuous integral equation leads to the MoM discretization (2.27).

Applying the operator $\mathcal{P}$ to an eigenfunction of the EFIE operator leads to

$$\mathcal{P}|u_q\rangle = \sum_{m=1}^{N} \langle t_m|u_q\rangle |t_m\rangle \tag{3.21}$$

Since u_q is a Fourier basis function, we can use the Fourier shift theorem to obtain

$$\langle t_m|u_q\rangle = e^{-jq\phi_m} T_{-q} \tag{3.22}$$

From the explicit expression given above for the eigenfunctions, we also have that
$e^{-jq\phi_m} = u_q(\phi_m)$. Combining these expressions gives

$$\mathcal{P}\,|u_q\rangle = \sum_{m=1}^{N} u_q(\phi_m)\,T_{-q}\,|t_m\rangle \tag{3.23}$$

This expression shows that when the Fourier function u_q is projected into the testing subspace using (3.20), the resulting expansion is given by samples of u_q at the locations of the testing functions, with an additional scale factor T_{-q}. This is the projection error associated with the testing functions.

When the EFIE operator is discretized, the eigenfunctions associated with ϕ and ϕ' in the cylindrical wave expansion (3.1) are projected into the testing and expansion subspaces, respectively, using the same projection (3.20) that led to (3.23). Projection of the ϕ and ϕ' dependence of the operator into the testing and expansion subspaces leads to scaling coefficients T_{-q} and F_q for each eigenfunction. These coefficients appear in the moment matrix eigenvalue (3.15) as factors in the first term.

Aliasing error.   Since the order $q + sN$ of the eigenvalue in the summation over $s = \pm1, \pm2, \ldots$ is always N or larger, the second term of (3.17) is determined by the high-order, unmodeled eigenvalues of the EFIE operator. To understand this contribution to the spectral error, we can consider the right-hand side of (3.6) to be a continuous function with respect to ϕ_m and ϕ_n. When ϕ_m and ϕ_n are sampled at discrete points, aliasing occurs for modes in the summation over q with $|q| > N/2$, since these modes oscillate more rapidly than the mesh Nyquist frequency. The eigenvalues associated with these aliased modes add to the eigenvalues of modeled modes. The spectral shift caused by this effect is aliasing error.

The eigenvalues of high-order, aliased modes are linked to the singularity of the Green's function or operator kernel. If the kernel were smooth, the eigenvalues would decay rapidly and the aliasing error would be small. The stronger the singularity of the kernel, the slower the decay of the eigenvalues as the order becomes large, and the larger the aliasing error.

The basis functions also have an effect on aliasing error. While (3.1) is singular at $\phi = \phi'$, the right-hand side of (3.6) is not singular at $\phi_m = \phi_n$ because of the regularizing effect of the expansion and testing functions. The basis functions ameliorate the singularity of the kernel and cause a more rapid falloff of the terms in the summation over q. The smoother the basis functions, the more rapidly the coefficients T_q and F_q fall off with increasing q. Thus, smooth basis functions reduce aliasing error. If the basis functions were so smooth as to be spatially bandlimited (e.g., sinc functions or eigenfunctions of the integral operator), then the aliasing error term would vanish, since bandlimited basis functions are orthogonal to the unmodeled modes with order $q > N/2$. We will see in Chapter 8 that higher-order polynomial basis functions act much like bandlimited functions and reduce aliasing error significantly relative to low-order basis functions.

From a physical point of view, to achieve high accuracy with the method of moments, the basis functions should properly model the electromagnetic interactions between modes with spatial frequencies below the mesh Nyquist frequency. If the basis functions have spectral content beyond the mesh Nyquist frequency, error increases. With nonsmooth basis functions, the singularities associated with discontinuities of the expansion functions radiate quasistatic fields that are received by the testing functions and perturb the self-interaction of the modeled modes, causing solution error.

Finally, we observe that because the aliasing error is an additive term, whereas the projection error is multiplicative, the effects of the two error contributions on the current and scattering amplitude errors will be different in character.

3.1.4 Spectral Error for Low-Order Basis Functions

We now develop explicit results for the spectral error with low-order, piecewise polynomial expansion and testing functions. These include the pulse function (2.30b) and triangle function (2.30c). We will also include the delta function (2.30a), which can be considered to be a basis function with smoothness one order smaller than that of the pulse function. These basis functions can be classified by polynomial order, so that the pulse function has order $p = 0$, the triangle function has order $p = 1$, and the delta function is $p = -1$.

These basis functions are members of a family of splines generated by convolutions of the pulse function [3]. The triangle function is the twofold convolution of the pulse function. If the width of the pulse function is h, then the width of the resulting triangle function is $2h$. The triangle function is identical to the finite element basis set generated by linear shape functions on a one-dimensional mesh. This family of splines can be extended further. The threefold convolution of a pulse function is a piecewise quadratic function of width $3h$. For higher orders, this type of spline approaches a Gaussian shape with a very large width of support, but still with only one basis function associated with each mesh element. In practice, basis functions of this class are rarely used beyond the triangle function ($p = 1$). Higher-order basis functions are typically constructed differently, by utilizing multiple polynomials on each mesh element, so that the basis set is complete up to order p (i.e., by taking linear combinations of the basis functions on a mesh element, any polynomial of order up to p can be represented exactly). We will consider higher-order basis functions in Chapter 8.

For the piecewise polynomial basis functions (2.30), the Fourier series coefficients T_q of the testing function are

$$T_q = \left[\frac{\sin(\pi q/N)}{\pi q/N} \right]^{p+1} \tag{3.24}$$

which is a sinc function raised to a power determined by the polynomial order of the testing function. The Fourier coefficients of the expansion function are similar, but with exponent determined by the polynomial order of the expansion function:

$$F_q = \left[\frac{\sin (\pi q/N)}{\pi q/N} \right]^{p'+1} \tag{3.25}$$

For delta testing functions (also called point testing or point matching), the exponent in (3.24) is zero and the Fourier series coefficients are all equal to one. For the pulse expansion function, (3.25) is a sinc function, and for the triangle function (3.25) is a squared sinc function. Since this class of low-order basis functions are convolutions of the pulse function in real space, it is to be expected that the Fourier series coefficients are products of sinc functions.

It can be seen from (3.17) that the spectral error depends on the product $T_{-q}F_q$. A consequence of this is that the spectral error is symmetric with respect to exchange of the testing and expansion functions (we will see later that this symmetry extends to the scattering amplitude solution error, but not to the current error). By forming the product of (3.24) and (3.25), it can be seen that the spectral error depends only on the sum of the polynomial orders of the testing and expansion functions. Accordingly, we define the parameter

$$b = p + p' + 2 \tag{3.26}$$

where p and p' are the polynomial orders of the testing and expansion functions.

With these results in hand, we are prepared to evaluate the spectral error in closed form for specific choices of basis functions. By making use of the asymptotic expansion (3.3) in (3.18), the relative spectral error can be approximated as

$$E_q \simeq T_{-q}F_q - 1 + \frac{j\eta}{2n_\lambda \lambda_q} \sum_{s \neq 0} \frac{T_{-q-sN}F_{q+sN}}{|s+q/N|} \tag{3.27}$$

Inserting the Fourier series coefficients (3.24) and (3.25) leads to

$$E_{q,b} \simeq \left[\frac{\sin (\pi \beta_q/n_\lambda)}{\pi \beta_q/n_\lambda} \right]^{b} - 1 + \frac{j\eta}{2n_\lambda \lambda_q} \sum_{s \neq 0} \frac{\mathrm{sgn}(s) \sin^b \pi(s + \beta_q/n_\lambda)}{\pi^b (s + \beta_q/n_\lambda)^{b+1}} \tag{3.28}$$

where $\beta_q = q/(k_0 a)$ is the normalized spatial frequency associated with the qth current mode. β_q is defined such that the mode with spatial frequency equal to k_0 corresponds to $\beta_q = 1$. Accordingly, $|\beta_q| < 1$ corresponds to radiating modes, and $|\beta_q| > 1$ to nonradiating modes.

Delta testing and expansion functions ($b = 0$). It is possible to obtain convergent spectral error with single-point evaluations of the operator kernel, as long as the testing and expansion points are different. As the analysis of this discretization scheme is simpler for a flat scatterer, discussion of the $b = 0$ case is deferred until Section 5.1.5.

Pulse expansion and point testing ($b = 1$). The simplest basis set that is applied to the TM-EFIE in common practice consists of pulse expansion functions and point testing. For this basis, the order parameter is $b = 1$. By expanding (3.28) for small β_q/n_λ and evaluating the summation, it can be shown that the relative spectral error is

$$E_{q,1} \simeq -\frac{\pi^2 \beta_q^2}{6n_\lambda^2} + \frac{1.8 j \eta \beta_q^2}{n_\lambda^3 \lambda_q} \qquad (3.29)$$

The exact constant in the second term is $3\zeta(3)/2$, where $\zeta(x)$ is the Riemann zeta function. The first term of this expression is the projection error associated with the qth eigenvalue, and the second term is due to aliasing of high-order modes. It can be seen that the projection error is second order in n_λ, whereas the aliasing error is third order. This will have ramifications later in determining the orders of the current and scattering amplitude solution errors.

Triangle expansion and point testing ($b = 2$). For point testing and triangle expansion functions or pulse testing and expansion functions ($b = 2$), the spectral error is

$$E_{q,2} \simeq -\frac{\pi^2 \beta_q^2}{3n_\lambda^2} + \frac{1.2 j \eta \beta_q^2}{n_\lambda^3 \lambda_q} \qquad (3.30)$$

where the exact constant in the second term is $\zeta(3)$. The spectral error for this basis set is identical to that obtained with pulse expansion and pulse testing functions, since $E_{q,b}$ only depends on the combined order parameter b and not on p and p' individually. The aliasing error term is third order and the projection error is second order.

For both pulse and triangle expansion functions, the projection error is second order (i.e., decreases as n_λ^{-2} as the mesh is refined) and the aliasing error is third order. Both error terms include a factor of β_q^2. As the spatial frequency of the eigenmode increases, the difficulty of representing the mode in the approximation space defined by the basis functions becomes greater, so it is natural that the spectral error associated with projection of the integral operator into the approximation space increases with the mode order q. The goal now is to relate the spectral error to current solution and scattering amplitude errors. Given that the spectral error consists of both second- and third-order terms, a natural question is, which of these terms has a dominant influence on the current and scattering amplitude errors?

3.1.5 Current Solution Error

The analysis above has quantified the error in the moment matrix, in terms of the spectral error or eigenvalue perturbation caused by discretization. To determine the current error, we will need to consider the effect of discretization on the incident field. If the incident field is a plane wave traveling along the x-axis, then the right-hand side of the linear system (2.27) has elements given by

$$E_n^{\text{inc}} = \int d\phi\, t_n(\phi) e^{-jk_0 a \cos \phi} \tag{3.31}$$

Using the cylindrical mode expansion of a plane wave, this can be written as

$$E_n^{\text{inc}} = \sum_{q=-\infty}^{\infty} j^{-q} J_q(k_0 a) T_{-q} e^{-jq\phi_n} \tag{3.32}$$

Each term in this sum is an eigenvector of the moment matrix $\mathbf{Z}$, so it is straightforward to apply $\mathbf{Z}^{-1}$ and obtain the vector of unknown coefficients of the numerical current solution:

$$\hat{J}_n = \sum_{q=-\infty}^{\infty} \frac{j^{-q} J_q(k_0 a) T_{-q}}{\hat{\lambda}_q} e^{-jq\phi_n} \tag{3.33}$$

In order to determine the solution error, we need a similar modal expansion for the exact current on the cylinder. From the Mie series solution for the circular cylinder, the exact current evaluated at the mesh element centers ϕ_n on the cylinder surface is

$$J_n = \sum_{q=-\infty}^{\infty} \frac{j^{-q} J_q(k_0 a)}{\lambda_q} e^{-jq\phi_n} \tag{3.34}$$

With the method of moment solution (3.33) and the exact current solution (3.34), we can compute the current solution error. To avoid the interpolation error introduced by the expansion functions and focus on error intrinsic to the current unknowns in (2.25), we will evaluate the error at the mesh element centers ϕ_n. Subtracting (3.33) from (3.34) leads to the current error

$$\Delta J_n = J_n - \hat{J}_n = -\frac{2}{\eta \pi k_0 a} \sum_{q=-\infty}^{\infty} \frac{j^{-q}(1 + E_q - T_{-q})}{H_q^{(2)}(k_0 a)(1 + E_q)} e^{-jq\phi_n} \tag{3.35}$$

The RMS current error is

$$\|\Delta J\|_{\text{RMS}} = \frac{2}{\eta \pi k_0 a} \left[\sum_{q=-\infty}^{\infty} \left| \frac{E_q^{(2)} + T_{-q}(F_q - 1)}{H_q^{(2)}(k_0 a)(1 + E_q)} \right|^2 \right]^{1/2} \tag{3.36}$$

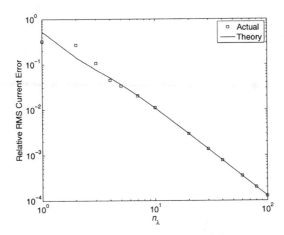

Figure 3.2: Relative RMS surface current solution error for a circular cylinder with radius $k_0 a = \pi$ and a TM-polarized incident field. The expansion functions are piecewise constant (pulse functions) and point testing is used. The MoM implementation is ideal, so that there is no geometrical discretization error, quadrature error, or linear system solution error. From the slope of the error curves, it can be seen that the current error is second order.

where $E_q^{(2)}$ is the aliasing error term of (3.18), which is

$$E_q^{(2)} = \frac{1}{\lambda_q} \sum_{s \neq 0} \lambda_{q+sN} T_{-q-sN} F_{q+sN} \tag{3.37}$$

A comparison of this result with numerical experiments is shown in Figure 3.2.

As $|q|$ becomes large relative to $k_0 a$, the Hankel function in the denominator of (3.36) becomes large, so that the dominant contribution to the sum comes from terms with small q. This follows from the fact that the current on the surface of the scatterer is smooth, and hence, the high spectral components are small. Hence, large q terms contribute little to the current error.

Since T_q and F_q are both close to unity for $|q| \leq N/2$, to leading order $T_{-q}(F_q - 1) \simeq F_q - 1$. As a consequence, the projection error associated with the testing functions cancels and does not reduce the current error (3.36). This can be seen more simply by ignoring the aliasing error and approximating (3.15) by the first term, and upon substituting $\hat{\lambda}_q$ into (3.33) the factor T_{-q} cancels. This occurs because the left- and right-hand sides of the EFIE in (2.8) are tested using the same set of functions, and the scaling effect of the testing functions cancels for each mode. We will see later that this partial cancellation of the projection error becomes complete for the scattering amplitude, and only aliasing error enters into the scattering amplitude error.

It is helpful to develop a simple, closed-form approximation for the RMS current error (3.36). Were it not for the problem of internal resonances, this would be straightforward. In the case of a flat PEC strip, which has no internal resonances since it is an open geometry, we will see later that the current error neglecting edge effects can be readily obtained in closed form. As long as the frequency of the incident wave is not close to an internal resonance of the cylinder, however, it can be found empirically from (3.36) that the relative RMS current error for point testing and pulse expansion functions can be approximated as [4]

$$\frac{\|\Delta J\|_{\text{RMS}}}{\|J\|_{\text{RMS}}} \simeq 0.7 n_\lambda^{-2} \tag{3.38}$$

As a function of the scatterer size $k_0 a$, this result is a lower bound on the current error, since the error increases dramatically in peaks near internal resonance frequencies. While the constant in this expression requires some numerical work to compute, the exponent on n_λ is ultimately more important and can be readily inferred from (3.25) and (3.36). For large n_λ, (3.25) goes as $1 - c_1 n_\lambda^{-2}$, which leads to a contribution of order n_λ^{-2} in (3.36). Since $E_q^{(2)}$ is third order and smaller, the second-order contribution dominates, and the overall current error has order n_λ^{-2}.

As noted above, we have used the RMS error measure to avoid including the large interpolation error caused by the pulse basis functions in the current error estimate. The L^2 error for the current solution is dominated by interpolation error, which leads to a much larger first-order error that is essentially identical to (2.62). One lesson from this is that if a continuous current approximation is needed, the current samples obtained with low-order expansion functions can be interpolated using higher-order polynomials to obtain a smoother and more accurate solution. The current unknowns resulting from a simple pulse basis implementation of MoM, for example, could be used as expansion coefficients with triangle functions, which linearly interpolate the current samples and the discontinuities in the current solution at the pulse function edges are eliminated.

Finally, it is interesting to consider these results in light of the rule of thumb, "ten unknowns per wavelength." At $n_\lambda = 10$, the relative RMS current error is

$$\frac{\|\Delta J\|_{\text{RMS}}}{\|J\|_{\text{RMS}}} \simeq 0.01 \tag{3.39}$$

or a 1% relative error. This is a very small error, due to the ideal implementation of the method of moments assumed in this section. If the idealizing assumptions about the discretization scheme do not hold, we will see in Section 3.3 that error can increase substantially.

3.1.6 Scattering Amplitude Error

Often in numerical simulations it is not the surface current itself that is the desired final result, but rather a derived quantity such as an impedance or far field. Accuracy for impedances is closely related to the accuracy of the current itself, but for far field computations, the transformation from currents to radiated fields can have a substantial effect on numerical accuracy. In this section, we will analyze the solution error for far fields in terms of the scattering amplitude.

Using (2.45), the numerical scattering amplitude solution can be written in terms of modal expansions of incident and scattered plane waves as

$$\hat{S}(\phi^{\text{sca}}, \phi^{\text{inc}}) = -\frac{k_0 \eta}{4} \sum_{m=1}^{N} \sum_{q,r} \frac{a_q^* b_r}{\hat{\lambda}_r} e^{jq(\phi^{\text{sca}} - \phi_m)} e^{-jr(\phi^{\text{inc}} - \phi_m)} \tag{3.40}$$

where a_q and b_q are the Fourier coefficients of the discretized scattered plane wave (2.43) and the incident plane wave (3.32), respectively. The Fourier coefficients of the scattered plane wave are similar to those of the incident field in (3.32), except that the scattered plane wave is discretized using the expansion functions instead of the testing functions, so T_q is replaced by F_q. Inserting the Fourier coefficients leads to

$$\hat{S} = -\frac{k_0}{4} \sum_{q,r} \left[hj^q J_q(k_0 a) F_q \right]^* \frac{1}{\hat{\lambda}_r} j^r J_r(k_0 a) T_{-r} e^{jq\phi^{\text{sca}} - jr\phi^{\text{inc}}} \sum_{m=1}^{N} e^{j(r-q)\phi_m} \tag{3.41}$$

Using (3.10), the sum over m is periodic with N, but due to the rapid growth of the eigenvalue $\hat{\lambda}_r$ with r, only the term for which $q = r$ is significant. Applying this to (3.41) and using $N = 2\pi a/h$ leads to

$$\hat{S}(\phi) = \frac{j\pi k_0 a}{2} \sum_{q=-\infty}^{\infty} \frac{J_q(k_0 a)^2 F_q T_{-q}}{\hat{\lambda}_q} e^{jq\phi} \tag{3.42}$$

where $\phi = \phi^{\text{sca}} - \phi^{\text{inc}}$. In terms of the relative spectral error E_q, this becomes

$$\hat{S}(\phi) = \frac{j\pi k_0 a}{2} \sum_{q=-\infty}^{\infty} \frac{J_q(k_0 a)^2 F_q T_{-q}}{\lambda_q (1 + E_q)} e^{jq\phi} \tag{3.43}$$

With (3.2) for the exact operator eigenvalue, we have

$$\hat{S}(\phi) = -\sum_{q=-\infty}^{\infty} \frac{J_q(k_0 a) F_q T_{-q}}{H_q^{(2)}(k_0 a)(1 + E_q)} e^{jq\phi} \tag{3.44}$$

The exact scattering amplitude is

$$S(\phi) = -\sum_{q=-\infty}^{\infty} \frac{J_q(k_0 a)}{H_q^{(2)}(k_0 a)} e^{jq\phi} \tag{3.45}$$

Subtracting the numerical and exact scattering amplitudes leads to the error

$$\Delta\hat{S}(\phi) = \sum_{q=-\infty}^{\infty} \frac{J_q(k_0 a)}{H_q^{(2)}(k_0 a)} \frac{E_q - (F_q T_{-q} - 1)}{1 + E_q} e^{jq\phi} \tag{3.46}$$

Using (3.18), the scattering amplitude error becomes

$$\Delta S(\phi) = -\sum_{q=-\infty}^{\infty} \frac{J_q(k_0 a)}{H_q^{(2)}(k_0 a)} \frac{E_q^{(2)}}{1 + E_q} e^{jq\phi} \tag{3.47}$$

In this expression, $E_q^{(2)}$ is the aliasing error component of the spectral error given by (3.37).

Instead of the absolute error (3.47), it is convenient to consider the scattering amplitude error in a relative sense. For point testing and pulse expansion functions, by numerical evaluation of (3.47) and (3.45) the relative backscattering amplitude error can be approximated as [4]

$$\frac{|\Delta S|}{|S|} \simeq 1.9(k_0 a)^{-1} n_\lambda^{-3} \tag{3.48}$$

This expression is valid away from the internal resonance frequencies of the scatterer. As with the surface current, the scattering amplitude error increases near internal resonances frequencies. It can be seen that the scattering amplitude error is third order, and decreases as n_λ^{-3} or h^3 as the mesh is refined. While this result was derived for the backscattering amplitude, the error for bistatic scattering amplitudes has the same order as the backscattering amplitude. Numerical results with a comparison to the theoretical error estimate are shown in Figure 3.3.

Inspection of (3.47) shows that the scattering amplitude error exhibits a very significant behavior with respect to its dependence on the spectral error. The projection error component of the spectral error has essentially no effect on the scattering amplitude. Only the aliasing error component $E_q^{(2)}$ appears in the numerator, rather than the full spectral error E_q. Typically, $E_q \ll 1$, so that the E_q term in the denominator can be neglected. To see this more simply, if we ignore the aliasing error and approximate (3.15) by the first term, then the factor $F_q T_{-q}$ cancels in (3.42).

For the low-order basis functions considered in this chapter, the spectral error consists of two terms, one of which is third order with respect to mesh density for the

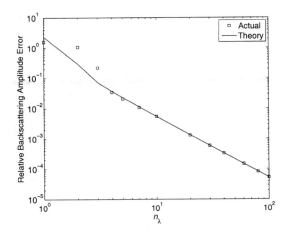

Figure 3.3: Relative backscattering amplitude error for a circular cylinder with radius $k_0 a = \pi$ and a TM-polarized incident field. The expansion functions are piecewise constant (pulse functions) and point testing is used. The MoM implementation is ideal, so that there is no geometrical discretization error, quadrature error, or linear system solution error. From the slope of the error curves, it can be seen that the backscattering amplitude converges at a third-order rate.

particular discretization scheme considered in this section, and another which is second order. The larger of these spectral error contributions, the projection error, does not contribute to the scattering amplitude error. As a result, the scattering amplitude is influenced only by the third-order aliasing error term and therefore has a higher order of convergence than the current. This cancellation of the projection error is closely related to a fundamental aspect of the method of moments, the variational property.

3.2 VARIATIONAL PRINCIPLES, THE MOMENT METHOD, AND SUPERCONVERGENCE

A variational principle is a functional for which the stationary point is the solution to a differential or integral equation. In electromagnetic theory, the functional typically takes a current solution as an argument and yields a scattering amplitude. The method of moments can be derived from a variational principle using the Rayleigh-Ritz procedure, by substituting an expansion for the current solution in terms of basis functions into the functional and setting the derivatives of the functional with respect to the expansion coefficients to zero.

In the computational electromagnetics community, variationality has been the subject of much consternation and numerous misunderstandings. Since a functional is stationary about the exact solution, its numerical value should not vary significantly from the exact value if it is evaluated for an approximate current solution. On this basis, it was suggested decades ago that a variational expression for the scattering amplitude should be used instead of the standard formula to improve the accuracy of the computed scattering amplitude. Jones showed that this is not the case, since the value of the functional is identical to that obtained by using the current solution obtained from Galerkin's method in the standard formula for the scattering amplitude [5]. Because of the close connection between Galerkin's method and variational formulas, it was further contended by some that Galerkin's method provides more accurate scattering amplitudes than the method of moments with different testing and expansion functions. Mautz [6] and Peterson, et al. [7] cleared up this misconception by demonstrating that the improved accuracy associated with Galerkin's method is obtained with the more general moment method, as long as the scattered plane wave used in computing the scattering amplitude is discretized using the expansion functions. Most recently, Dudley observed that if an integral operator is poorly behaved, there is no reason to expect good solution convergence just because a numerical solution can be derived from a variational principle [8].

Variational principles for the EFIE can be given for 2D and 3D problems. We will consider the 3D EFIE operator here, but all the results hold in very similar form for the TM-EFIE and TE-EFIE. For the EFIE, the variational principle is obtained from the bivariate functional [7]

$$I(\mathbf{J}_s, \mathbf{J}_s^a) = \langle \mathbf{E}^s, \mathbf{J}_s \rangle + \langle \mathbf{J}_s^a, \mathbf{E}^{\text{inc}} \rangle - \langle \mathbf{J}_s^a, \mathcal{T}\mathbf{J}_s \rangle \tag{3.49}$$

The angle brackets denote the vector L^2 inner product on the scatterer surface (in [7], the angle brackets denote the symmetric product without a complex conjugate). $\mathbf{J}^a$ is the solution to the adjoint problem

$$\hat{n} \times \mathcal{T}^a \mathbf{J}_s^a = \hat{n} \times \mathbf{E}^s \tag{3.50}$$

where $\mathbf{E}^s$ is a plane wave propagating away from the scatterer and the adoint operator is defined by

$$\langle \mathbf{f}_1, \mathcal{T}\mathbf{f}_2 \rangle = \langle \mathcal{T}^a \mathbf{f}_1, \mathbf{f}_2 \rangle \tag{3.51}$$

for all pairs of vector fields $\mathbf{f}_1$ and $\mathbf{f}_2$ on the scatterer surface. Interestingly, the adjoint operator is not needed in most applications of the variational principle, but it can be constructed if needed using the definition (3.51). Since $\mathcal{T}^a \mathbf{J}_s^a$ is equal to the tangential part of the scattered plane wave $\mathbf{E}^s$ on the scatterer, we can use (3.51) to show that

$$\langle \mathbf{E}^s, \mathbf{J}_s \rangle = \langle \mathcal{T}^a \mathbf{J}_s^a, \mathbf{J}_s \rangle = \langle \mathbf{J}_s^a, \mathcal{T}\mathbf{J}_s \rangle = \langle \mathbf{J}_s^a, \mathbf{E}^{\text{inc}} \rangle \tag{3.52}$$

As shown in Section 2.6, if the incident field is a plane wave, then $\langle \mathbf{E}^s, \mathbf{J}_s \rangle$ is proportional to the scattering amplitude in the direction of the plane wave $\mathbf{E}^s$. From these equalities, it follows that if J_s and J_s^a are exact solutions to the EFIE (2.7) and adjoint problem (3.50), respectively, then the functional (3.49) is equal (up to a constant factor) to the scattering amplitude.

Using the calculus of variations, the EFIE can be obtained from the functional. If we require the functional to be stationary with respect to J_s, so that

$$\delta_{J_s} I(J_s, J_s^a) = 0 \tag{3.53}$$

where $\delta_{J_s} I(J_s, J_s^a)$ is the first variation of the functional with respect to J_s, then it can be shown that J_s satisfies the EFIE (2.7). By inserting the approximate solution (2.31) and a similar approximation

$$\hat{J}_s^a = \sum_{n=1}^{N} I_n^a t_n \tag{3.54}$$

for the adjoint solution J_s^a expanded in terms of the testing functions, we can apply the Rayleigh-Ritz procedure to the functional. Minimizing the functional with respect to the coefficients I_n in the approximation (2.31) leads to the same linear system

$$\sum_{n=1}^{N} \langle t_m, \mathcal{T} f_n \rangle I_n = \langle t_m, E^{\text{inc}} \rangle \tag{3.55}$$

that is obtained with the method of moments [7]. For well behaved operators, the Rayleigh-Ritz procedure finds the solution in the approximation space for which the value of the functional is closest to the exact value.

In Section 2.4, the method of moments linear system was derived using the method of weighted residuals. That approach is ad hoc in the sense that there is no guarantee that the method will lead to accurate numerical results. The variational approach, on the other hand, can be used to gain additional information about the accuracy of a computed value for the functional (3.49). Ignoring an unimportant constant, the scattering amplitude error can be expressed as

$$\begin{aligned}
\Delta S &= \langle E^s, \Delta J_s \rangle \\
&= \langle J_s^a, \mathcal{T} \Delta J_s \rangle \\
&= \langle \Delta J_s^a, \mathcal{T} \Delta J_s \rangle + \langle \hat{J}_s^a, \mathcal{T} \Delta J_s \rangle
\end{aligned} \tag{3.56}$$

By inserting the approximation (3.54) for the adjoint solution in terms of testing functions t_n and assuming that $\hat{J}_s$ in (2.31) is obtained from the method of moments, so that the linear system (3.55) is satisfied, it is straightforward to demonstrate that the second term on the right-hand side vanishes. From this result, it follows that the scattering amplitude error is [8]

$$\Delta S = \langle \Delta J_s^a, \mathcal{T} \Delta J_s \rangle \tag{3.57}$$

At least formally, the scattering amplitude error is second order with respect to the current solution error. If the error ΔJ_s^a for the method of moments solution to the adjoint problem (3.50) is similar in magnitude to the solution error ΔJ_s for the EFIE, the scattering amplitude error should be the square of the two small current errors.

Even though many of the misconceptions about variationality have been cleared up in recent years, the error expression (3.57) is still poorly understood. One significant issue is that for low-order basis functions, the solution errors ΔJ_s and ΔJ_s^a are not continuous functions. If a quadrature rule is used to evaluate the moment matrix elements, for example, then the basis and testing functions are effectively delta functions, in which case the expansion (2.31) and the solution errors become distributions. As a consequence, it is difficult to use (3.57) to make rigorous conclusions about the actual accuracy of scattering amplitude solutions [7, 8].

3.2.1 Superconvergence

Despite the complications associated with the variational error formula (3.57), it is true that the accuracy of the scattering amplitude is often better than that of the current solution. We have already seen that the scattering amplitude error for an ideal discretization of the TM-EFIE with pulse basis functions and point matching is third order, while the RMS current error is second order. We will refer to this higher-order accuracy of the scattering amplitude relative to the current solution as superconvergence. What is surprising is that for other discretizations (as we will see shortly) the scattering amplitude is not superconvergent, even though the variational error expression (3.57) still holds.

Fortunately, the spectral series for the scattering amplitude error in (3.47) provides a more concrete explanation for the superconvergence of the scattering amplitude. As discussed above, the projection error term in the spectral error cancels and does not impact the scattering amplitude. When the incident and scattered fields are projected into the testing and expansion subspaces, each term in the modal expansion of the fields is scaled by a scale factor T_q associated with the testing functions for the incident field or a factor of F_q due to the expansion functions for the scattered field. Similarly, when the integral operator is discretized, each term in the modal expansion of the operator is scaled by the factor $T_q F_q$, since the operator is projected into both the testing and expansion subspaces. These scale factors are associated with the projection error term of the spectral error (3.18). When the moment matrix is inverted and the scattering amplitude formed according to (2.45), these scale factors cancel on a mode-by-mode basis.

As a result of this mode-by-mode cancellation, the projection error $T_{-q} F_q - 1$ does not appear in the scattering amplitude error (3.47). The projection error portion of the spectral error has no impact on the scattering amplitude, and the accuracy of the scattering amplitude is determined solely by the aliasing error. For the ideal discretization considered above, the aliasing error is third order and smaller than the second-order projection error, so the scattering amplitude error has a higher-order accuracy than the current error. This mode-by-mode cancellation of the projection error is the underlying mechanism by which superconvergence of the scattering amplitude is realized.

3.2.2 Idealizing Assumptions

One important aspect of the superconvergence property of the method of moments is that it typically only occurs for ideal implementations of the method of moments. The method of moments implementation considered previously in this chapter is optimistic in several respects, which we will briefly outline here. As we will see in the next section, relaxing these idealizing assumptions can destroy the superconvergence of the scattering amplitude.

One idealizing assumption that may not be obvious relates to the smoothness of the incident field. For a plane wave, the variation of the field along the scatterer is very smooth, and as a result, the cylindrical mode expansion coefficients decrease rapidly for high orders. This decay in the incident field expansion coefficients causes the terms in (3.36) to become exponentially small and negligible for nonpropagating modes ($|q| \geq k_0 a$), and so the current error is determined by the spectral error only for propagating modes ($|q| \leq k_0 a$). For a less smooth incident field, the expansion coefficients of the incident field decay more slowly, so the spectral error associated with higher-order modes cannot be neglected. Since the spectral error increases with order of the eigenvalue, the solution error also increases as the incident field becomes less smooth. If the incident field is produced by a line source located at a distance d from the cylinder, for example, it can be shown that the current error depends on the spectral error E_q for roughly $|q| \leq 2a/d$, which exceeds $k_0 a$ when the line source is closer than λ/π to the cylinder. If the incident field is nonsingular (i.e., the source does not lie on the scatterer), the increased error represents a larger error constant, but the convergence order of the solution remains the same.

Other idealizing assumptions relate to the implementation of the method of moments, rather than to the physics of the scattering problem. First, an exact geometrical model is used for the cylinder, so that the facets are curved, rather than flat. Second, the number of integration points for the moment matrix integrals is sufficiently large that the quadrature error (3.61) is negligible and the integrations are essentially exact. Finally, the linear system is inverted using a direct method, so the linear system solution error is negligible. We refer to this type of implementation as ideal because the solution error is associated only with projection of the continuous integral operator into a finite dimensional approximation subspace. We turn now to discretizations that relax the first two of these idealizing assumptions (the third, linear system solution error, we defer to Chapter 9).

3.3 TM-EFIE with Nonideal Discretizations

We now consider the accuracy of the method of moments for more practical implementations involving numerical integration error and flat-facet meshes. This involves relaxing the idealizing assumptions made in previous sections. The second-order accuracy of the current solution and third-order accuracy for scattering amplitudes for an ideal MoM discretization of the EFIE provides a baseline that can be used to quantify the effects of nonideal implementations. The analysis of the previous sections for an idealized MoM implementation represents unavoidable error that is associated with projection of the EFIE onto the finite dimensional subspace spanned by a particular set of basis functions. In practice, nonideal effects can significantly increase the solution error (although in some special cases, such as a single-point quadrature rule, a nonideal discretization can actually have better accuracy).

As in the previous section, we will begin by determining the spectral error caused by nonideal discretizations. The results in the previous section for current and scattering amplitude error in terms of spectral error will then be used to relate the spectral error to solution errors.

3.3.1 Quadrature Error

In this section, we will relax the assumption of exact integration of moment matrix elements, and introduce into the analysis the concept of quadrature error. The effect of approximate numerical integration can be taken into account by replacing the continuous integral in the Fourier transform (3.8) of the expansion function $f(\phi)$ with the quadrature rule used to evaluate moment matrix elements, so that F_q becomes

$$F_{q,M} = \frac{1}{\theta_0} \sum_{n=1}^{M} w_n f(\xi_n) e^{jq\xi_n} \tag{3.58}$$

where M is the order of the quadrature rule and w_n are the weights corresponding to the integration points or abscissas ξ_n. The Fourier series coefficients T_q of the testing function $t(\phi)$ is modified similarly. With these changes to the coefficients F_q and T_q, the expression (3.27) for the spectral error incorporates the effect of integration error. In effect, we have replaced the basis and testing functions with new basis functions consisting of combinations of delta functions located at the integration points associated with the quadrature rule. These effective basis functions are not as smooth as pulse or triangle functions, so in general approximate numerical integration causes solution error to increase.

To proceed with the analysis, we will consider a specific integration rule. For the M-point Riemann sum with the integrand evaluated at the centers of subintervals (the midpoint or Euler quadrature rule), the weights are $w_n = \delta = \theta_0/M$ and the abscissas are $\xi_n = (n - 1/2)\delta - \theta_0/2$. In the case of piecewise constant expansion functions $(p' = 0)$, $F_{q,M}$ becomes the periodic sinc or Dirichlet function

$$F_{q,M} = \frac{\sin \frac{\pi q}{N}}{M \sin \frac{\pi q}{MN}} \tag{3.59}$$

This function has maxima at $q = MNr$, where $r = 0, \pm 1, \pm 2, \ldots$. The maxima dominate the summation over s in (3.27), so to evaluate the spectral error we can simply retain the maxima, which for small q occur at $s = Mr$, and neglect all other terms in the sum. At the maxima, $F_{q+sN,M} \simeq (-1)^{r(M+1)}$. Inserting this into (3.27) leads to

$$E_{q,M}^{(2)} \simeq \frac{j\eta}{2n_\lambda \lambda_q} \sum_{r \neq 0} \frac{(-1)^{r(M+1)}}{|Mr|} \tag{3.60}$$

If M is even, this reduces to the alternating harmonic series $\sum_{k=1}^{\infty} (-1)^{k-1}/k = \ln 2$ (for odd M, one of the integration points lies exactly on the singularity of the Green's function, and the sum is infinite). The aliasing error becomes

$$E_{q,M}^{(2)} \simeq -\frac{j\eta \ln 2}{Mn_\lambda \lambda_q} \tag{3.61}$$

for small β_q. The quadrature rule also has a small effect on the projection error term of (3.18), but the additional contribution is of the same order as the projection error for exact integration and so is less important than the aliasing error contribution in (3.61).

The most significant aspect of the quadrature error (3.61) is that it is first order in n_λ^{-1}. Since the quadrature error is lower order than other spectral error contributions, it becomes dominant as the mesh is refined. As we will see shortly, this makes both the current and scattering amplitude errors first order. In short, quadrature error dramatically decreases the accuracy of the method of moments.

We will pause briefly to gain some intuition into the large increase in error caused by numerical integration. As observed in Section 3.1.3, the aliasing error term in the spectral error is associated with the smoothness of the basis functions—the smoother the basis set, the smaller the aliasing error. High-order eigenmodes associated with quasistatic fields are not excited by smooth basis functions and so do not perturb the interactions of the modeled eigenmodes having spatial frequency within the "Nyquist" range for a given mesh. The effect of an integration rule is to make the expansion and testing functions effectively linear combinations of delta functions. The effectively nonsmooth basis functions strongly excite high-order eigenmodes, so that in general aliasing error is much larger than it would be with exactly integrated, smooth basis functions.

The above analysis was for the eigenvalue perturbation introduced by approximate integration, but from (3.36) and (3.47) it can be seen that quadrature error has a direct impact on the current and scattering amplitude errors. Since quadrature error contributes a large, first-order error term to each eigenvalue, by inspection of (3.36) and (3.47) the current and scattering amplitude errors also become first order. The only question is the value of the constants in the error estimates. In the solution error expressions, the quadrature error appears in a sum over eigenmodes, which for a plane wave incident field can be readily approximated, yielding simple closed form approximations for the relative RMS current error [4]

$$\frac{\|\Delta J\|_{\text{RMS}}}{\|J\|_{\text{RMS}}} \simeq \frac{1}{Mn_\lambda} \tag{3.62}$$

and the backscattering amplitude error,

$$\frac{|\Delta S|}{|S|} \simeq \frac{1}{Mn_\lambda} \tag{3.63}$$

Unlike the ideal discretization analyzed in Section 3.1.6, for which the scattering error was two orders of magnitude better than the current error, with quadrature error taken into consideration the scattering amplitude and current errors both converge at only a first-order rate. The superconvergence of the scattering amplitude is destroyed by the error associated with approximate integration.

Assuming $M = 5$ (or the single-point rule with analytical integration of diagonal moment matrix elements discussed below), ten mesh elements per wavelength now corresponds to a relative error of approximately 2% for both the current and scattering amplitude. This is larger than would be the case with exact integration, and the error increases at a much slower rate as the mesh is refined.

3.3.2 Reducing Quadrature Error

There are several ways to reduce quadrature error and restore the higher-order convergence of the method of moments. The simplest (but not necessarily the most computationally efficient) is to increase the number of integration points, so that the value of M in (3.61) is larger and the quadrature error is smaller. This postpones the onset of first-order error convergence to a larger value of n_λ. If the mesh is refined, eventually the first-order quadrature error term will become dominant. If the quadrature rule is sufficiently accurate, the reduction to first order can be delayed to a fine enough mesh that for practical purposes, the convergence rate of the numerical method may be considered to be better than first order.

The problem with this approach is that the number of quadrature points must be very large to ensure that quadrature error is smaller than other spectral error terms, leading to a high computational cost to evaluate moment matrix elements. Typically, however, a higher-order quadrature rule is only required for singular and near-singular moment matrix element integrations. This effectively increases the value of M in (3.61) without requiring as many integration points for interactions between widely separated testing and expansion functions. To improve efficiency further, singularity subtraction can be used to analytically integrate the leading singularity of the integrand for overlapping testing and expansion functions, leaving a less singular integral to be evaluated by numerical quadrature.

Another approach is to choose the weights and abscissas of the integration rule judiciously so as to decrease the magnitude of the quadrature error. For the simple midpoint rule analyzed here, the only freedom available is the number of integration points M, which must be sufficiently large in order for the quadrature error not to dominate the solution error. More sophisticated integration rules, either based on nonclassical Gaussian quadrature or coordinate transformations that render the integrand nonsingular, can significantly reduce quadrature error. The computational electromagnetics community has devoted much effort to the development of these kinds of improved quadrature rules [9, 10].

As a simple model for quadrature error reduction, we consider briefly the common discretization for which off-diagonal elements of the moment matrix are integrated using a single integration point $(M = 1)$, and the diagonal elements or self-interaction terms are evaluated by expansion of the kernel and integrating the singularity analytically. It can be shown that the aliasing error has the same form as (3.61), but with a smaller constant than would be obtained if the midpoint rule were used for the diagonal moment matrix elements [11]:

$$E_{q,1}^{(2)} \simeq -\frac{j\eta}{n_\lambda \lambda_q} (\ln \pi - 1) \tag{3.64}$$

By comparing this with (3.61), it can be seen that the spectral error for this integration scheme is equivalent to the use of the midpoint rule for all moment matrix elements with five integration points per mesh element $(M = 5)$.

3.3.3 Geometrical Discretization Error

Another common practical deviation from the ideal method of moments implementation is a flat-facet mesh representation of the scatterer geometry. Due to the complexity of generating conformal meshes and implementing a method of moments algorithm with basis functions defined on curved mesh elements, software packages relying on simpler flat-facet implementations remain in routine use. The goal of this section is to understand the impact of flat mesh elements on solution error.

To determine the effect of a flat-facet mesh, we use the modal series (3.1) in a more general form allowing the source and field points to lie off the scatterer surface. Be-

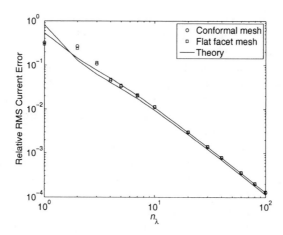

Figure 3.4: Relative RMS surface current solution error for conformal and flat-facet meshes. Integral equation: EFIE. Polarization: TM. Basis: point testing, pulse expansion functions. Moment matrix element integration: exact. Scatterer: circular cylinder, radius $k_0 a = \pi$.

cause the deviation of the flat-facet mesh from the true scatterer profile is small, the terms in the series can be expanded about $\rho = a$. This leads to an additional term in the spectral error due to geometrical discretization. For point testing and pulse expansion functions, the geometrical discretization error is [4]

$$E_q^{(3)} \simeq \frac{\eta \pi^3}{6 k_0 a n_\lambda^2 \lambda_q} G_q \tag{3.65}$$

where

$$G_q = J_q(k_0 a) H_q^{(2)}(k_0 a) + \frac{k_0 a}{2} J_q(k_0 a) H_q^{(2)\prime}(k_0 a) + \frac{j}{2\pi} \tag{3.66}$$

The geometrical discretization error contribution can be seen to be second order.

Since the leading contribution to the current error with a conformal mesh is also second order, the convergence rate of the current solution is neither increased nor decreased by geometrical discretization error, but the absolute error is larger. The RMS current error can be found by combining (3.65) with the other spectral error contributions in (3.36). Empirically, it can be observed that away from internal resonance frequencies the RMS current solution error is

$$\frac{\|\Delta J\|_{\mathrm{RMS}}}{\|J\|_{\mathrm{RMS}}} \simeq n_\lambda^{-2} \tag{3.67}$$

From this expression, it can be seen that the order of the error is the same as for an ideal discretization, and the impact of geometrical discretization error on the current

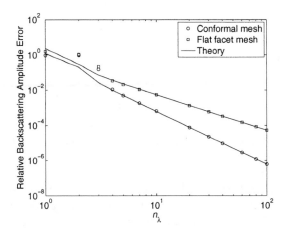

Figure 3.5: Relative backscattering amplitude error for conformal and flat-facet meshes. Integral equation: EFIE. Polarization: TM. Basis: point testing, pulse expansion functions. Moment matrix element integration: exact. Scatterer: circular cylinder, radius $k_0 a = \pi$. Geometrical discretization error reduces the convergence rate from third to second order.

solution for the TM-EFIE is small. A comparison of numerical and theoretical errors is shown in Figure 3.4.

We now want to determine the scattering amplitude error. In general, the flat-facet mesh affects the discretization of the incident and scattered plane waves used to compute the scattering amplitude, as well as the integral operator. For the current, the flat-facet mesh had no impact on the incident field, because point testing was used. Since (2.46) is discretized with the expansion functions, the flat-facet mesh leads to an additional error contribution through the scattered plane wave. Combining the two contributions leads to the estimate

$$\frac{|\Delta S|}{|S|} \simeq 1.6 \, (k_0 a)^{-1} n_\lambda^{-2} \tag{3.68}$$

for the relative backscattering amplitude error. A comparison of numerical results and the error estimate is shown in Figure 3.5.

With the current, geometrical discretization increases the magnitude of the error, but the order of convergence remains the same. For the scattering amplitude, on the other hand, the order of convergence decreases from third for the ideal implementation of the method of moments to second order. While the degradation of the accuracy is not as severe as for the case of quadrature error, the order of the scattering amplitude is no longer better than that of the current solution. From this, it follows that geometrical discretization error destroys the superconvergence property of the method of moments.

3.4 TE-EFIE

With respect to the TM-EFIE, the chief difference of the EFIE for the TE polarization is the stronger singularity of the kernel in the operator (2.11). Because of the presence of derivatives in the definition, $\mathcal{N}$ is an integro-differential operator. For the TM-EFIE, even though the kernel of the operator is singular, the singularity is weak enough that the operator has a smoothing effect, so that it maps a given surface current to a smoother tangential electric field. From the point of view of the operator spectrum, the high-order eigenvalues of the TM-EFIE operator decay asymptotically, as can be seen from (3.4). Consequently, when the integral operator acts on highly oscillatory components of a given function, the amplitude of these components is reduced and the function becomes smoother.

The TE-EFIE, on the other hand, has the opposite behavior. Due to the derivatives in the operator, the singularity of the kernel is stronger than for the TM polarization, so $\mathcal{N}$ is referred to as a hypersingular operator. We will see shortly that the eigenvalues of the operator increase as the order becomes large. Highly oscillatory components in a surface current are amplified by the TE-EFIE operator, leading to an "antismoothing" or differentiating effect. The main ramification for the method of moments is that smoother basis functions are required to obtain stable numerical results with the TE-EFIE.

3.4.1 Spectral Error

As with the TM polarization, the first step in the analysis of the TE case is to obtain a spectral decomposition of the discretized operator for a circular cylinder scatterer geometry. Using the expansion (3.1), the moment matrix elements associated with the operator $\mathcal{N}$ in (2.11) can be expressed as

$$Z_{mn} = \frac{\eta k_0 a}{4\theta_0} \int \int d\phi \, d\phi' \, t_m(\phi) \sum_q J_q(k_0 a) H_q^{(2)}(k_0 a)$$

$$\times \left[\cos(\phi - \phi') f_n(\phi') - \frac{jq}{(k_0 a)^2} \frac{\partial f_n(\phi')}{\partial \phi'} \right] e^{-jq(\phi - \phi')} \quad (3.69)$$

Expanding $\cos(\phi - \phi')$ into exponentials, integrating the second term by parts, and making use of recursion relations for the derivatives of the Bessel and Hankel functions yields

$$Z_{mn} = \frac{\eta \pi k_0 a}{2} \sum_q J_q'(k_0 a) H_q^{(2)\prime}(k_0 a) T_{-q} F_q e^{-jq(\phi_m - \phi_n)} \quad (3.70)$$

where as before T_{-q} and F_q are the Fourier series coefficients of the testing and expansion functions, respectively. By proceeding as in Section 3.1, the eigenvalues of the moment matrix are found to be

$$\hat{\lambda}_q = \frac{\eta \pi k_0 a}{2} \sum_{s=-\infty}^{\infty} J'_{q+sN}(k_0 a) H^{(2)'}_{q+sN}(k_0 a) T_{-q-sN} F_{q+sN} \qquad (3.71)$$

This expression is identical to (3.12), except that the Bessel and Hankel functions are replaced with their first derivatives.

By taking the limit of (3.71) as $N \to \infty$, it can be seen that the eigenvalues of the continuous TE-EFIE operator are

$$\lambda_q = \frac{\eta \pi k_0 a}{2} J'_q(k_0 a) H^{(2)'}_q(k_0 a) \qquad (3.72)$$

The relative spectral error of (3.71) with respect to the exact eigenvalue is

$$E_q \simeq T_{-q} F_q - 1 - \frac{j \eta n_\lambda}{2 \lambda_q} \sum_{s \neq 0} |s + q/N| \, T_{-q-sN} F_{q+sN} \qquad (3.73)$$

where we have used the asymptotic expansion

$$J'_\nu(x) H^{(2)'}_\nu(x) \sim -\frac{j|\nu|}{\pi x^2}, \quad \nu \to \infty \qquad (3.74)$$

From this expression, it can be seen that the eigenvalues grow linearly in magnitude with the order, in contrast to (3.4), which decays inversely with order.

As with the TM polarization, the spectral error consists of two contributions, one due to high-order eigenvalues associated with the singularity of the kernel of the integral operator (aliasing error), and another due to projection of eigenmodes onto the basis subspace (projection error). In terms of the spectral error, the current and scattering amplitude errors have the same forms as (3.36) and (3.47) for the TM polarization, but with the Bessel and Hankel functions replaced by their first derivatives.

3.4.2 Spectral Error for Low-Order Basis Functions

The antismoothing property of the TE operator is manifested in the growth of the high-order eigenvalues, which increase linearly as the order becomes large. To compensate for this, in general from (3.73) it can be seen that the Fourier series coefficients of the testing and expansion functions must decrease more rapidly with order, indicating that smoother basis functions are needed.

The case of pulse expansion functions with point testing (smoothness index $b = 1$) is an exception to this and actually leads at least in some cases to convergent numerical results. For pulse functions, the series in (3.73) is not absolutely convergent, as the magnitudes of the terms in the series do not become small as q becomes large. Due to the behavior of the phase of the terms in the series, however, the sum is actually finite and the spectral error converges with respect to n_λ.

For pulse expansion functions, to obtain the spectral error for small β_q/n_λ analytically, we must employ an additional term of the expansion $J_\nu'(x)H_\nu^{(2)'}(x) \sim i|\nu|/(\pi x^2) + jx^2/(2\pi|\nu|)$ in (3.71). This leads to the result

$$E_{q,1} \simeq -\frac{\pi^2\beta_q^2}{6n_\lambda^2} - \frac{0.9\,j\eta\beta_q^2}{n_\lambda^3\lambda_q} \tag{3.75}$$

Surprisingly, the spectral error converges to zero as n_λ becomes large, which indicates that convergent numerical solutions can actually be obtained with point testing and pulse expansion functions. Triangle expansion functions with point testing can also provide convergent results, but the spectral error is divergent if the testing point is located at the apex of the triangle functions, so the testing point must be shifted away from the mesh element centers (see Section 5.1.5).

Because the TE-EFIE operator includes derivatives, one might expect that smoother basis functions would be required to obtain meaningful results. Even though the pulse and triangle expansion functions with point testing are in principle not sufficiently smooth for discretization of the TE-EFIE, convergent results can still be obtained due to cancellation of singularities in the moment matrix interactions. These discretizations are unstable, however, because the analysis assumes that the mesh is regular and testing points are precisely located to achieve the required cancellations. Physically, the large static fields radiated by the delta function charge distributions at the edges of the pulse functions exactly cancel at the center of the pulse functions, so testing at the pulse centers avoids the error associated with the discontinuities of the basis functions. For other scatterers or for irregular meshes, this cancellation may not occur.

To ensure a robust discretization, the basis functions must be sufficiently smooth that the series in (3.73) is absolutely convergent [12]. Asymptotically, the eigenvalues of the operator increase in magnitude, so that $\lambda_q \sim q$, $q \to \infty$. For absolute convergence of the spectral error series, the product of the Fourier series coefficients of the testing and expansion functions must decay at least as $1/q^3$, corresponding to a smoothness index of $b = 3$. A basis set with this degree of smoothness is pulse testing $(1/q)$ combined with triangle expansion functions $(1/q^2)$. Another possibility is point testing combined with piecewise quadratic expansion functions. The former is more commonly used in practice.

For the $b = 3$ discretization (pulse testing and triangle expansion functions, or point testing with piecewise quadratic expansion functions) , the spectral error is [11]

$$E_{q,3} \simeq -\frac{\pi^2\beta_q^2}{2n_\lambda^2} - \frac{1.8\,j\eta\beta_q^4}{n_\lambda^3\lambda_q} \tag{3.76}$$

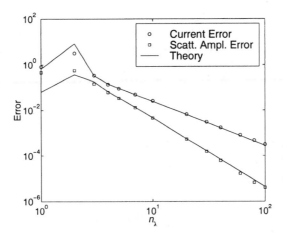

Figure 3.6: Relative RMS surface current error and backscattering amplitude error for a circular cylinder for the EFIE, TE polarization, and an ideal implementation of the method of moments. Basis: pulse testing, triangle expansion functions ($b = 3$). Moment matrix element integration: exact. Scatterer: circular cylinder, radius $k_0 a = \pi$. (©2005 IEEE [13].)

for small β_q/n_λ. Numerical results for the $b = 3$ discretization are shown in Figure 3.6. The surface current error is determined by the projection error term of the spectral error, which is second order in the mesh density. As with the TM-EFIE, the variational property of the scattering amplitude leads to a cancellation of the leading term in the spectral error, and the scattering amplitude converges at a third-order rate with respect to mesh density.

3.4.3 Quadrature Error

As for the TM polarization, the ideal discretization in the analysis leading up to (3.76) yields an optimistic, best-case error estimate when compared to practical implementations of the method of moments. If moment matrix integrals are not evaluated exactly, error increases. We will analyze the case of pulse testing and triangle expansion functions ($b = 3$) with approximate integration using the M-point integration rule described in Section 3.3.1.

The hypersingular term of (2.11) is typically integrated by parts to reduce the singularity of the kernel before application of a numerical quadrature rule. Integration by parts transfers the derivative operators to the basis functions. Since the derivative of the pulse function is a delta function doublet, the testing integration for the hypersingular term of the integral operator can be evaluated analytically and no quadrature rule is required. The integration over the source point includes the derivative of a triangle function, which is a pulse doublet.

The quadrature error caused by the source integration can be analyzed using the approach of Section 3.3.1. The basis function in (3.58) becomes a pulse doublet. Evaluating the sum analytically for the midpoint quadrature rule leads to

$$F'_{q,M} = -\frac{2j \sin \frac{\pi q}{N}}{\theta_0 \sin \frac{\pi q}{MN}} \tag{3.77}$$

Since the pulse doublet is an odd function, the Fourier series coefficients are imaginary. We will also require the Fourier series coefficients of the derivative of the pulse function, which are given by

$$T'_{q,M} = -2j\theta_0^{-1} \sin \frac{\pi q}{N} \tag{3.78}$$

After integration by parts, the kernel of the hypersingular term reduces to the weakly singular kernel of the TM polarization, and the quadrature error term of the spectral error becomes

$$E^{(2)}_{q,M} = -\frac{\eta \pi}{2k_0 a \lambda_q} \sum_{s \neq 0} J_{q+sN}(k_0 a) H^{(2)}_{q+sN}(k_0 a) T'_{-q-sN,M} F'_{q+sN,M} \tag{3.79}$$

Expanding the Bessel and Hankel functions asymptotically and evaluating the sum over s leads to

$$E^{(2)}_{q,M} \simeq \frac{j\eta \ln 2\beta_q^2}{n_\lambda M \lambda_q} \tag{3.80}$$

The quadrature error is first order with respect to mesh density, and has the same order as (3.61) for the weakly singular kernel. This result only includes the quadrature error for the hypersingular term of the operator, but since the weakly singular term is similar to the TM operator, the total spectral error can be approximated by the sum of (3.61) and (3.80). As with the TM polarization, quadrature error destroys the superconvergence of the method of moments, and the surface current and scattering amplitude errors are both first order.

3.4.4 Geometrical Discretization Error

The behavior of the TE-EFIE with respect to geometrical discretization error is similar to that shown in Figures 3.4 and 3.5 for the TM polarization. With a flat-facet mesh, a second-order error term is introduced into the spectral error. As a result, the scattering amplitude convergence rate is reduced from third to second order. Since the convergence rate of the current solution is already second order, the order of the current error does not change, but there is a small increase in the absolute error.

3.5 Solution Error for Other Smooth Scatterers

After a long treatment of error analysis for a circular PEC cylinder, one naturally wonders whether solution errors for other scatterers behave similarly, or if changes in the scatterer geometry can radically change the accuracy of the method of moments. Asymptotic error estimates like (1.1) are general and apply to a large class of geometries, but the constants in the estimates are unknown and the predicted convergence rate is only a lower bound. It seems clear that the more quantitative error estimates derived in this chapter must depend on the scatterer geometry. Circular cylinder results certainly cannot be applied to nonsmooth geometries with edges and corners, so nonsmooth scatterers will be considered separately in Chapter 5.

Even for smooth geometries, it is not obvious how the circular cylinder results might generalize to other scatterers. As observed at the beginning of this chapter, the circular cylinder has special properties that are not general for smooth scatterers. The EFIE operator for the circular cylinder is normal, meaning that the surface current solution can be decomposed into a set of orthogonal operator eigenfunctions, whereas for other geometries, current eigenfunctions exist but are not orthogonal and the operator is nonnormal. It is not known precisely how operator nonnormality affects the numerical accuracy of the moment method, although it appears that the degree of nonnormality at least in some cases is relatively weak [11].

On the other hand, the behavior of a discretized solution on a relatively smooth region of the scatterer might be expected to have a similar projection error, since this depends primarily on local properties of the basis functions rather than the global scatterer geometry. Aliasing error depends on the operator kernel singularity, which is also a localized function. By these arguments, accuracy might be expected to be largely independent of the large-scale geometry of the scatterer, making the error estimates obtained above valid for a wide class of smooth scatterers.

The relationship of circular cylinder error estimates to solution error for other scatterer geometries must be considered in two parts. First, do the solution convergence rates such as the third-order rate reflected by the estimate (3.48) for the scattering amplitude with an ideal discretization of the EFIE extend to other scatterers? Second, if the convergence rates are the same, is the constant in the solution error larger or smaller than that of the circular cylinder?

These questions can be studied using the empirical approach of [4]. In that work, eight scatterers with parametric equations given in Table 3.1 were used to test solution error for a variety of smooth geometries. The scatterer cross sections are shown in Figure 3.7. As exact solutions are not available for the noncircular geometries, reference solutions were generated with a very fine mesh and a method of moments implementation with third-order solution convergence. The scatterers were illuminated

Table 3.1: Parametric equations for smooth scatterer cross-sections. The first scatterer is the circular cylinder. Figure 3.7 shows the geometries of the scatterers. The markers in the third column are used in Figure 3.8.

	$x(t)$	$y(t)$	
1	$1.1698 \cos t$	$1.1698 \sin t$	•
2	$0.2588 \cos t(4 + \cos 3t)$	$0.2588 \sin t(4 + \cos 3t)$	○
3	$1.5173 \cos t$	$0.7586 \sin t$	×
4	$1.4237 \sin (\cos 2t)$	$1.4237 \cos (-2\cos t + \sin t)$	+
5	$1.2715 \sin [\cos (t + 1)]$	$1.2715 \cos t$	*
6	$0.2040 \cos t(4 + \cos 6t)$	$0.2040 \sin t(4 + \cos 6t)$	□
7	$1.1963 \cos (t + 0.5 \cos t)$	$1.1963 \sin (t + 0.5 \sin t)$	▷
8	$0.7586 \cos t$	$1.5173 \sin t$	▽
9	$-0.4659 \cos (2 + 3\cos t)$	$1.3977 \sin t$	△

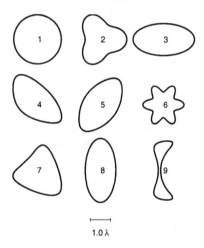

1.0 λ

Figure 3.7: Scatterer geometries corresponding to the parametric equations in Table 3.1. Each shape has a perimeter of approximately 7.4λ. (©2005 IEEE [4].)

by a plane wave incident from $0°$ with respect to the standard orientation in Figure 3.7. Numerical solution errors for these scatterers are shown in Figure 3.8.

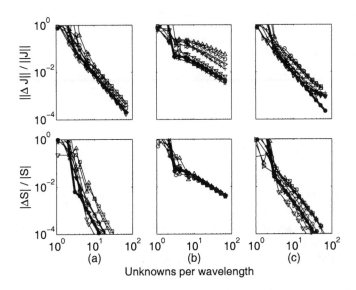

Figure 3.8: Relative RMS current error and backscattering amplitude error for the EFIE with TM polarization for the scatterers shown in Figure 3.7. (a) Ideal discretization. (b) Quadrature rule for moment matrix integrations. (c) Flat-facet mesh. Current convergence rates are second, first, and second order, respectively. Backscattering amplitude convergence rates are third, first, and second order, respectively. Larger errors correspond to scatterers with highly curved features. (©2005 IEEE [4].)

From the results in Figure 3.8, it can be seen that solution convergence rates for these scatterers are the same as that of the circular cylinder. For some of the scatterers, however, the absolute error is significantly larger. The scatterer geometries with high curvature features have larger error than the circular cylinder. In the limit as curvature becomes infinite, a sharp corner results, and the current solution is singular at the corner. Because the basis functions cannot represent a singular or near-singular current as accurately as a smooth current, error increases. This effect will be studied further in Chapter 5.

In certain cases, with a flat-facet mesh, the EFIE convergence rate degrades and is not second order (Figure 3.8(c), upper plot). This is caused by the behavior of flat-facet mesh element generation near inflection points of the geometry. The order of the scattering amplitude error is unaffected by this localized mesh property.

3.6 SUMMARY

We have examined in this chapter the behavior of method of moment solutions for the EFIE in the case of a circular cylinder scatterer geometry. For low-order basis func-

Table 3.2: Solution error convergence rates for the method of moments for a TM-polarized plane wave incident on a circular PEC cylinder. Pulse expansion functions and point testing are used. $n_\lambda = \lambda/h$ is the mesh density, where h is the mesh element width. These estimates are lower bounds, since error can increase significantly near internal resonance frequencies.

	Ideal	Quadrature Rule	Flat-Facet Mesh
EFIE RMS Current Error	$0.7\, n_\lambda^{-2}$	$M^{-1} n_\lambda^{-1}$	n_λ^{-2}
EFIE Backscattering Error	$1.9\, (k_0 a)^{-1} n_\lambda^{-3}$	$M^{-1} n_\lambda^{-1}$	$1.6\, (k_0 a)^{-1} n_\lambda^{-2}$

tions and an ideal implementation of the method of moments, the convergence rate of the surface current solution, where the error is measured using RMS difference with respect to the exact solution at mesh node points, is second order. The convergence rate of the scattering amplitude is third order. The more rapid convergence of the scattering amplitude is a manifestation of the variational property of the moment method.

For nonideal discretizations, solution error increases and the superconvergence property is destroyed. If moment matrix elements are integrating using low-order numerical quadrature rules, the current and scattering amplitude convergence rates decrease to first order. If the mesh elements are flat, rather than conformal or curved to match the scatter surface exactly, the current error remains second order and the scattering amplitude convergence rate decreases from third to second order.

The error estimates obtained in this chapter are summarized in Table 3.2 for the TM polarization. Errors for the TE polarization have the same order in n_λ, but the constants in the error estimates may be different, and smoother basis functions are generally required. By checking empirical errors observed for a collection of smooth scatterers, it can be seen that error convergence rates for the circular cylinder are valid for other scatterers, but the absolute error increases for highly curved geometries.

For a mesh with the rule-of-thumb density of ten mesh elements per wavelength ($n_\lambda = 10$), it can be seen from the results in this chapter that error varies widely, depending on the details of the method of moments implementation. For an ideal discretization, relative errors of 1% or better for current and scattering amplitude errors can be expected. For a flat-facet mesh, errors at $n_\lambda = 10$ are similar in magnitude to the ideal case. If a low-order quadrature rule is used to evaluate moment matrix elements, accuracy worsens to 5%-10%.

References

[1] A. W. Naylor and G. R. Sell, *Linear Operator Theory in Engineering and Science*. New York: Springer-Verlag, 1982.

[2] A. G. Ramm, "Eigenfunction expansion of a discrete spectrum in diffraction problems," *Radiotek. i Elektron.*, vol. 18, pp. 364–369, 1973.

[3] A. G. Dallas, G. C. Hsiao, and R. E. Kleinman, "Observations on the numerical stability of the Galerkin method," *Adv. Comput. Math.*, vol. 9, pp. 37–67, 1998.

[4] C. P. Davis and K. F. Warnick, "Error analysis of 2D MoM for MFIE/EFIE/CFIE based on the circular cylinder," *IEEE Trans. Ant. Propag.*, vol. 53, pp. 321–331, Jan. 2005.

[5] D. S. Jones, "A critique of the variational method in scattering problems," *IRE Trans. Ant. Propag.*, vol. AP-4, no. 3, pp. 297–301, 1956.

[6] J. Mautz, "Variational aspects of the reaction in the method of moments," *IEEE Trans. Ant. Propag.*, vol. 42, pp. 1631–1638, Dec. 1994.

[7] A. F. Peterson, D. R. Wilton, and R. E. Jorgenson, "Variational nature of Galerkin and non-Galerkin moment method solutions," *IEEE Trans. Ant. Propag.*, vol. 44, April 1996.

[8] D. G. Dudley, "Comments on 'Variational nature of Galerkin and non-Galerkin moment method solutions,'" *IEEE Trans. Ant. Propag.*, vol. 45, June 1997.

[9] S. Wandzura, "Accuracy in computation of matrix elements of singular kernels," *11th Annual Review of Progress in Applied Computational Electromagnetics*, vol. II, Monterey, CA, pp. 1170–1176, Naval Postgraduate School, Mar. 20–25, 1995.

[10] M. A. Khayat and D. R. Wilton, "Numerical evaluation of singular and near-singular potential integrals," *IEEE Trans. Ant. Propag.*, vol. 53, pp. 3180–3190, Oct. 2005.

[11] K. F. Warnick and W. C. Chew, "On the spectrum of the electric field integral equation and the convergence of the moment method," *Int. J. Numer. Meth. Engr.*, vol. 51, pp. 31–56, May 2001.

[12] M. I. Aksun and R. Mittra, "Choices of expansion and testing functions for the method of moments applied to a class of electromagnetic problems," *IEEE Trans. Micr. Th. Tech.*, vol. 41, pp. 503–509, Mar. 1993.

[13] K. F. Warnick and W. C. Chew, "Accuracy of the method of moments for scattering by a cylinder," *IEEE Trans. Micr. Th. Tech.*, vol. 48, pp. 1652–1660, Oct. 2000.

Chapter 4

Error Analysis of the MFIE and CFIE

with Clayton P. Davis

Having laid the groundwork for error analysis of the EFIE in the previous chapter, we will now apply these techniques to the magnetic field integral equation (MFIE) and the combined field integral equation (CFIE). The focus will be on similarities in the numerical behavior of the method of moments with respect to the EFIE, as well as new phenomena unique to the MFIE. As before, we will first consider an ideal implementation, with an exact geometrical representation of the scatterer and exact integration of moment matrix elements. Quadrature error and geometrical discretization error will then be introduced into the analysis. We will then step back and consider all the integral formulations in terms of the operator smoothing properties, solution error convergence rates, and behavior with respect to the variational property of the method of moments.

The key differences of the MFIE with respect to the EFIE are:

> *Second-kind integral equation.* The MFIE includes an identity operator that strongly influences the numerical behavior the moment method.

> *Smooth kernel.* The kernel of the integral part of the MFIE operator is smoother than the kernel of the EFIE.

Due to the smoothness of the kernel, quadrature error is less significant for the MFIE than for the EFIE. Both the identity term and the integral part of the operator can be readily discretized with low-order integration rules. In spite of this, we will find that the presence of the identity operator in the second-kind integral equation reduces the accuracy of the method of moments for low-order basis functions. Higher-order basis functions or regularization of the operator kernel can be used to remedy this effect. As might be expected, solution error for the CFIE is determined by the larger of the errors for the EFIE and MFIE. In this chapter, we will focus on the TM polarization, since the behavior of the MFIE and CFIE for the TE polarization is similar to the TM case.

4.1 TM-MFIE WITH IDEAL DISCRETIZATIONS

For the circular cylinder, the MFIE operator can be represented using a spectral decomposition in terms of eigenvalues and eigenfunctions. As with the EFIE in the previous chapter, the spectral decomposition can be used to analyze the error of numerical solutions to the MFIE using the method of moments [1]. The eigenfunctions of the operator are Fourier functions of the form $e^{jq\phi}$, $q = 0, \pm 1, \pm 2, \dots$. The eigenvalues of the operator $\mathcal{I}/2 + \mathcal{M}_{TM}$ defined in (2.18) are

$$\lambda_q^{\mathcal{M}_{TM}} = 1 - \tfrac{1}{2} j\pi k_0 a J_q(k_0 a) H_q^{(2)\prime}(k_0 a) = -\tfrac{1}{2} j\pi k_0 a J_q'(k_0 a) H_q^{(2)}(k_0 a) \qquad (4.1)$$

where the two forms for the eigenvalue are related by the Wronskian identity for Bessel functions. Using the eigenvalues and eigenfunctions, the MFIE operator can be represented in spectral form as

$$\left(\tfrac{1}{2}\mathcal{I} + \mathcal{M}_{TM}\right) u(\phi) = \tfrac{1}{2} u(\phi) + \frac{1}{2\pi} \sum_{q=-\infty}^{\infty} \left(\lambda_q^{\mathcal{M}_{TM}} - \tfrac{1}{2}\right) e^{-jq\phi} \int_0^{2\pi} d\phi' e^{jq\phi'} u(\phi')$$

$$(4.2)$$

The integral term is an inner product of u with the eigenfunction $e^{jq\phi}$. The inner product selects the qth Fourier series coefficient of u, which is then scaled by the qth operator eigenvalue. The existence of a spectral representation of the operator of this form indicates that the MFIE, like the EFIE, is a normal operator for the circular cylinder.

4.1.1 Operator Smoothing Properties

To obtain an accurate numerical solution with the method of moments, generally the basis functions used to expand the surface current J should possess the same properties that are characteristic of J. Piecewise linear basis functions mimic the continuity of the exact current J, while second-order basis functions would also provide continuity of charge. The smoothness of a basis function can be quantified in terms of the falloff of the Fourier coefficients F_q defined in (3.8)—the smoother the basis function, the more quickly the coefficients F_q decay as q becomes large.

Besides the basis functions themselves, we have seen in Chapter 3 that there is another notion of smoothness related to the method of moments: the smoothing property of the integral operator. This was first encountered in explaining the differences between the TM-EFIE and TE-EFIE integral equations. The smoothness property of an operator reveals itself in the asymptotic growth or decay of the eigenvalues with respect to order. For the TM-EFIE, (3.4) shows that the eigenvalues of $\mathcal{L}$ decay as q^{-1}, and the operator decreases the amplitudes of rapidly oscillating components of the function that it acts on. This operator is a low-pass or integrating operator. For the TE-EFIE, the eigenvalues of $\mathcal{N}$ grow linearly with order, and the operator is a differentiating or antismoothing operator.

How does the MFIE fit into this classification? To answer this question, we must consider the two terms of the operator separately. The eigenvalue (4.1) can be expanded asymptotically using high-order approximations for the Bessel and Hankel functions to obtain

$$\lambda_q^{\mathcal{M}_{\text{TM}}} \sim \tfrac{1}{2} + \frac{(k_0 a)^2}{4|q|^3}, \quad q \to \infty \tag{4.3}$$

The second term is the asymptotic expansion of the eigenvalues of the integral part $\mathcal{M}_{\text{TM}}$ of the MFIE alone. Since the decay rate is q^{-3}, the operator has an even stronger smoothing behavior than the TM-EFIE operator $\mathcal{L}$. The same holds for the TE-MFIE and $\mathcal{M}_{\text{TE}}$. Using the theory of Sobolev spaces, these smoothing considerations can be extended to noncircular scatterers as well [2]. In general, the operators $\mathcal{M}_{\text{TM}}$ and $\mathcal{M}_{\text{TE}}$ have a range space that is smoother in the Sobolev sense than the domain.

For the identity term of the MFIE, the eigenvalues are identically $1/2$, and this part of the MFIE operator is of course neither smoothing nor differentiating. As can be seen from (4.3), the identity operator eigenvalues dominate at large orders, and the full MFIE operators $\mathcal{I}/2 + \mathcal{M}_{\text{TM}}$ and $\mathcal{I}/2 + \mathcal{M}_{\text{TE}}$ are also neither smoothing nor differentiating (for scatterers with singularities such as corners or edges, this reasoning breaks down). The smoothing properties of both the identity and integral parts of the operator will have important ramifications for the accuracy of moment method solutions.

4.1.2 Discretized Operator Spectrum

We now want to determine the eigenvalues of the moment matrix associated with the MFIE operator. We will take the testing and expansion functions t_n and f_n and the corresponding Fourier coefficients T_q and F_q to be defined as in Section 3.1.1. Applying (2.28) to (4.2) and using the definitions of T_q and F_q, the moment matrix elements are given by

$$Z_{mn} = \tfrac{1}{2} R_{mn} + \frac{1}{N} \sum_q \left(\lambda_q^{\mathcal{M}_{\text{TM}}} - \tfrac{1}{2} \right) T_{-q} F_q e^{-jq(\phi_m - \phi_n)} \tag{4.4}$$

where ϕ_m are the mesh element center points as defined in Section 3.1.1. R_{mn} represents a matrix of overlap integrals of the testing and expansion functions given by

$$R_{mn} = \frac{1}{\theta_0} \int t_m(\phi) f_n(\phi) \, d\phi \tag{4.5}$$

For local basis functions, the overlap matrix is sparse with nonzero elements only near the main diagonal.

At this point, we will make the assumption that the basis functions are orthogonal, so that $R_{mn} = \delta_{mn}$. This property holds for low-order basis sets such as pulse expansion functions with point testing. Although it seems innocuous, the assumption of orthogonality significantly impacts the accuracy of the method of moments for the MFIE, as will be discussed further in Section 4.5.

Obtaining the eigenvalues of the moment matrix from (4.4) is similar to the derivation of (3.13) from (3.9), except that care must be taken when manipulating the infinite sums that a nonconvergent series does not occur along the way and lead to an incorrect result. For the MFIE, this procedure results in

$$\hat{\lambda}_q^{\mathcal{M}_{\mathrm{TM}}} = \frac{1}{2} + \sum_{s=-\infty}^{+\infty} \left(\lambda_{q+sN}^{\mathcal{M}_{\mathrm{TM}}} - \frac{1}{2} \right) T_{-q-sN} F_{q+sN} \tag{4.6}$$

The index q lies in the range $(-N/2, N/2)$, again ignoring unimportant complications associated with N odd, and the functions F_q and T_q are Fourier series coefficients of the expansion and testing functions as defined in (3.7) and (3.8). For terms of the sum over s in (4.6) with $s \neq 0$, we can use the asymptotic expansion (4.3) to obtain

$$\hat{\lambda}_q^{\mathcal{M}_{\mathrm{TM}}} = \frac{1}{2} + F_q T_q \left(\lambda_q^{\mathcal{M}_{\mathrm{TM}}} - \frac{1}{2} \right) + \frac{(k_0 a)^2}{4} \sum_{s \neq 0} \frac{F_{q+sN} T_{-q-sN}}{|q + sN|^3} \tag{4.7}$$

It is apparent that this expression contains terms similar to the projection and aliasing error contributions defined in Section 3.1.3.

4.1.3 Spectral Error

The relative spectral error associated with the moment matrix eigenvalue relative to the exact operator eigenvalue is

$$E_q = \frac{\hat{\lambda}_q^{\mathcal{M}_{\mathrm{TM}}} - \lambda_q^{\mathcal{M}_{\mathrm{TM}}}}{\lambda_q^{\mathcal{M}_{\mathrm{TM}}}} \tag{4.8}$$

Using (4.7) in this expression leads to

$$E_q \simeq \frac{\lambda_q^{\mathcal{M}_{\mathrm{TM}}} - \frac{1}{2}}{\lambda_q^{\mathcal{M}_{\mathrm{TM}}}} \left(F_q T_{-q} - 1 \right) + \frac{(k_0 a)^2}{4 \lambda_q^{\mathcal{M}_{\mathrm{TM}}}} \sum_{s \neq 0} \frac{F_{q+sN} T_{-q-sN}}{|q + sN|^3} \tag{4.9}$$

Using the terminology of Section 3.1.3, the first term is projection error, given by

$$E_q^{(1)} = \frac{\lambda_q^{\mathcal{M}_{\mathrm{TM}}} - \frac{1}{2}}{\lambda_q^{\mathcal{M}_{\mathrm{TM}}}} \left(F_q T_{-q} - 1 \right) \tag{4.10}$$

The second term is the aliasing error,

$$E_q^{(2)} = \frac{(k_0 a)^2}{4\lambda_q^{\mathcal{M}_{\mathrm{TM}}}} \sum_{s \neq 0} \frac{F_{q+sN} T_{-q-sN}}{|q + sN|^3} \tag{4.11}$$

Equation (4.10) differs from the projection error (3.18) for the EFIE in a subtle but important way. The leading factor of $(\lambda_q^{\mathcal{M}_{\mathrm{TM}}} - 1/2)/\lambda_q^{\mathcal{M}_{\mathrm{TM}}}$ means that there is no projection error associated with the identity term of the MFIE operator $\mathcal{I}/2 + \mathcal{M}_{\mathrm{TM}}$, since the identity operator discretizes to an identity matrix, without any mode scaling by the Fourier coefficients of the basis functions. This is a consequence of the assumption of orthogonality of the expansion and testing function. As it will turn out in Section 4.1.6, the *lack* of projection error associated with the identity operator actually increases the scattering amplitude error and causes a failure of the superconvergence property for the MFIE with low-order, orthogonal basis functions.

4.1.4 Current Solution Error

In order to determine the current solution error for the MFIE, we need to consider the discretization of the right-hand side of the integral equation, which is the magnetic field associated with the incident wave. For the circular cylinder, the tangential component of the incident magnetic field on the scatterer surface is given by

$$H_t^{\mathrm{inc}}(\boldsymbol{\rho}) = \frac{1}{\eta} e^{j\mathbf{k}^{\mathrm{inc}} \cdot \boldsymbol{\rho}} \cos\left(\phi^{\mathrm{inc}} - \Omega\right) \tag{4.12}$$

where Ω is the angle of the surface normal at $\boldsymbol{\rho}$ measured from the positive x-axis. Since the spectral error obtained above is given in terms of operator eigenfunctions, we must also expand the incident field in terms of eigenfunctions. Using the cylindrical wave expansion of a plane wave given by (3.32), together with the recursion relation for the derivative of a Bessel function, the elements of the right-hand side of the linear system (2.27) for the MFIE can be expressed as

$$b_m = \frac{1}{\eta} \sum_{q=-\infty}^{+\infty} j^{q-1} J_q'(k_0 a) T_q e^{-jq(\phi^{\mathrm{inc}} - \phi_m)} \tag{4.13}$$

The eigenvectors of the moment matrix are of the form

$$\mathbf{v}_q = \left[e^{-jq\phi_1}, \ e^{-jq\phi_2}, \dots, \ e^{-jq\phi_N} \right]^T \tag{4.14}$$

from which it can be seen that each term of the sum in (4.13) is an eigenvector. We can therefore apply $\mathbf{Z}^{-1}$ to $\mathbf{b}$ and find the unknown surface current expansion coefficients by dividing each term by the corresponding eigenvalue. This procedure leads to the result

$$I_n = \frac{1}{\eta} \sum_q \frac{j^{q-1} J_q'(k_0 a) T_q e^{-jq(\phi^{\text{inc}} - \phi_n)}}{\hat{\lambda}_q^{\mathcal{M}_{\text{TM}}}} \tag{4.15}$$

for the unknown coefficients in the numerical surface current expansion (2.25).

To find the current solution error, we must determine the difference between (4.15) and the corresponding values of the coefficients for the exact current on the scatterer surface. We will assume that the basis functions are interpolatory, so that $f_n(\phi_n) = 1$. In the limit as $N \to \infty$, (4.15) approaches the Mie series expansion for the exact current solution at the mesh element centers ϕ_n, which is

$$J(\phi_n) = \frac{2}{\pi k_0 a \eta} \sum_q \frac{j^q e^{-jq(\phi^{\text{inc}} - \phi_n)}}{H_q^{(2)}(k_0 a)} \tag{4.16}$$

The current solution error at the mesh node point ϕ_n is the difference of (4.15) and (4.16), which is

$$\Delta J_n = \frac{1}{\eta} \sum_q \frac{j^{q-1} J_q'(k_0 a) e^{-jq(\phi^{\text{inc}} - \phi_n)} (T_q - E_q - 1)}{\lambda_q^{\mathcal{M}_{\text{TM}}} (E_q + 1)} \tag{4.17}$$

This is a rather complicated expression and yields no obvious insight into the numerical behavior of the method of moments, other than the simple observation that current error depends on the spectral error. Fortunately, for a specific choice of basis functions we can simplify this result considerably and obtain a simple closed form approximation for the solution error.

By making use of the exact eigenvalue (4.1) in (4.17), the relative RMS current error can be expressed as

$$\frac{\|\Delta J\|_{RMS}}{\|J\|_{RMS}} = \|J\|_{RMS}^{-1} \frac{2}{\pi \eta k_0 a} \left[\sum_q \left| \frac{E_q + 1 - T_q}{H_q^{(2)}(k_0 a)(E_q + 1)} \right|^2 \right]^{1/2} \tag{4.18}$$

To simplify this expression, we can estimate the RMS value of the exact current by applying the definition of the RMS norm to the Mie series for the exact current solution. This leads to

$$\|J\|_{RMS} \simeq \frac{2}{\eta \pi k_0 a} \left[\sum_{q=-\infty}^{+\infty} \left| \frac{1}{H_q^{(2)}(k_0 a)} \right|^2 \right]^{1/2} \tag{4.19}$$

This expression is a function of $k_0 a$ only, and by numerical evaluation of the norm, the approximation

$$\|J\|_{RMS} \simeq \frac{2k_0 a + 1}{2\eta k_0 a} \tag{4.20}$$

can be developed. Substituting this into (4.18) yields

$$\frac{\|\Delta J\|_{RMS}}{\|J\|_{RMS}} \simeq \frac{4}{\pi(2k_0a+1)} \left[\sum_q \left| \frac{E_q+1-T_q}{H_q^{(2)}(k_0a)(E_q+1)} \right|^2 \right]^{1/2} \tag{4.21}$$

This expression is valid for any combination of orthogonal testing and expansion functions. To obtain a closed form approximation, we must make a particular choice for the basis functions.

4.1.5 Current Error for the Point/Pulse Discretization

For point testing and pulse expansion functions, from (3.24) and (3.25), $T_q = 1$ and $F_q = \text{sinc}(q/N)$. The projection error term (4.10) of the spectral error is dominant for this basis set. Inserting the Fourier coefficient of the expansion functions and expanding (4.10) for large n_λ yields

$$E_q^{(1)} \simeq -\frac{\lambda_q^{\mathcal{M}_{TM}} - \frac{1}{2}}{\lambda_q^{\mathcal{M}_{TM}}} \frac{\pi^2 q^2}{6(k_0a)^2 n_\lambda^2} \tag{4.22}$$

for the projection error. The aliasing error $E_q^{(2)}$ is fourth order in n_λ^{-1} and is negligible. This occurs because the kernel of the integral part $\mathcal{M}_{TM}$ of the operator $\mathcal{I}/2 + \mathcal{M}_{TM}$ is nonsingular. A singular kernel leads to a slower decay of high-order operator eigenvalues and large aliasing error, whereas a smooth kernel is associated with very small high-order eigenvalues and consequently the aliasing error is small.

Inserting the spectral error (4.22) into (4.21) leads to the current solution error estimate

$$\frac{\|\Delta J\|_{RMS}}{\|J\|_{RMS}} \simeq \frac{2\pi}{3(k_0a)^2(2k_0a+1)} \left[\sum_q \left| \frac{q^2\left(1-(2\lambda_q^{\mathcal{M}_{TM}})^{-1}\right)}{H_q^{(1)}(k_0a)} \right|^2 \right]^{1/2} n_\lambda^{-2} \tag{4.23}$$

where we have assumed that $E_q \ll 1$, which holds as long as the frequency is not near an internal resonance of the scatterer. The current error is second order with respect to the mesh element length.

As a final simplification to the current error estimate for the MFIE, we can determine the k_0a dependence numerically. Near internal resonances, error increases significantly, but away from internal resonances the dependence of the error on the scatterer size and frequency is relatively simple. This effect can be seen in Figure 4.1, which shows the RMS current error as a function of the scatterer electrical size. Away from internal resonances, the error is only weakly dependent on k_0a, and the error estimate

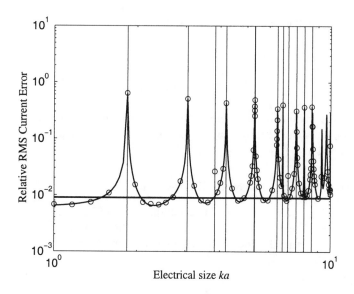

Figure 4.1: Relative RMS current error for the MFIE. Polarization: TM. Basis: point testing, pulse expansion functions. Moment matrix element integration: exact. Scatterer: circular cylinder. Mesh: conformal. Solid curve: theoretical error estimate (4.21). Horizontal line: approximation (4.24). Circles: numerical results for MoM current error. Vertical lines: internal resonance frequencies. (©2005 IEEE [1].)

can be simplified to

$$\frac{\|\Delta J\|_{RMS}}{\|J\|_{RMS}} \simeq 0.9\, n_\lambda^{-2} \tag{4.24}$$

At $k_0 a \simeq 3.83$ and $k_0 a \simeq 7.02$, neither this error estimate nor the more accurate error estimate (4.23) matches the observed numerical results. These points correspond to resonances of the $q = 0$ mode. This mode is constant along the scatterer surface, and the projection error for the corresponding eigenvalue is zero. When the $q = 0$ mode is internally resonant, because the projection error is zero the aliasing error, which was neglected in deriving (4.23), becomes significant and leads to a difference between the error estimate and actual numerical error. Otherwise, (4.24) provides a reasonable lower bound for the relative RMS current error. This error estimate is shown as a function of mesh density in Figure 4.2.

4.1.6 Scattering Amplitude Error

The numerical scattering amplitude solution for the MFIE can be determined from (3.40) using the Fourier coefficients of the discretized scattered plane wave (2.43) and the discretized incident field (4.13). Inserting the Fourier coefficients into (3.40) leads

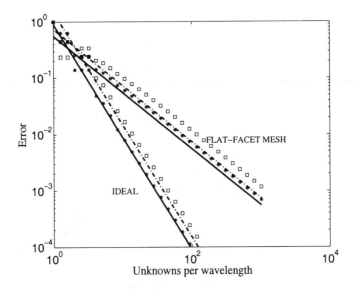

Figure 4.2: Solution errors for the TM-MFIE with ideal and flat-facet discretizations for a circular cylinder of electrical size $k_0 a \simeq 2.7$. Stars: MoM current error. Squares: MoM backscattering amplitude error. Solid lines: theoretical current error estimates from (4.24) (ideal) and (4.38) (flat-facet mesh). Dotted lines: backscattering error estimates from (4.31) (ideal) and (4.39) (flat-facet mesh). The RMS current errors and backscattering amplitude errors are first order for flat-facet implementations and second order for ideal implementations. (©2005 IEEE [1].)

to

$$\hat{S} = -\frac{k_0}{4} \sum_{q,r} \left[h j^q J_q(k_0 a) F_{-q} \right]^* \frac{1}{\hat{\lambda}_r^{\mathcal{M}_{\text{TM}}}} j^{r-1} J'_r(k_0 a) T_r e^{jq\phi^{\text{sca}} - jr\phi^{\text{inc}}} \sum_{m=1}^{N} e^{j(r-q)\phi_m}$$

(4.25)

The sum over m can be evaluated using (3.10) to obtain

$$\hat{S}(\phi) = \frac{j\pi k_0 a}{2} \sum_q \frac{J_q(k_0 a) J'_q(k_0 a) F_{-q} T_q}{\hat{\lambda}_q^{\mathcal{M}_{\text{TM}}}} e^{jq\phi}$$

(4.26)

where $\phi = \phi^{\text{sca}} - \phi^{\text{inc}}$. We can use (4.8) to express the numerical scattering amplitude in terms of the spectral error as

$$\hat{S}(\phi) = \frac{j\pi k_0 a}{2} \sum_q \frac{J_q(k_0 a) J'_q(k_0 a) F_{-q} T_q}{\lambda_q^{\mathcal{M}_{\text{TM}}} (1 + E_q)} e^{jq\phi}$$

(4.27)

Using (4.1) for the exact operator eigenvalue leads to

$$\hat{S}(\phi) = -\sum_q \frac{J_q(k_0 a) F_{-q} T_q}{H_q^{(2)}(k_0 a)(1 + E_q)} e^{jq\phi} \tag{4.28}$$

for the numerical scattering amplitude.

The exact scattering amplitude is given by (3.45). Subtracting the numerical and exact scattering amplitude leads to

$$\Delta\hat{S}(\phi) = \sum_q \frac{J_q(k_0 a)}{H_q^{(2)}(k_0 a)} \frac{E_q - (F_{-q} T_q - 1)}{1 + E_q} e^{jq\phi} \tag{4.29}$$

which is identical to (3.46) for the EFIE except that the relative spectral error E_q is different for the MFIE operator. By inserting the spectral error (4.9) in the numerator, we obtain the scattering amplitude error estimate

$$\Delta\hat{S}(\phi) = \sum_q \frac{J_q(k_0 a)}{H_q^{(2)}(k_0 a)} \frac{-\frac{1}{2\lambda_q^{M_{TM}}} E_q^{(1)} + E_q^{(2)}}{1 + E_q} e^{jq\phi} \tag{4.30}$$

Unlike the EFIE scattering amplitude error (3.47), for which only the aliasing error contributes to the error, in this expression both the projection and aliasing error are significant. This property and its ramifications will be discussed at length shortly.

4.1.7 Scattering Amplitude Error for the Point/Pulse Discretization

The scattering amplitude error can be simplified further by choosing a specific discretization scheme. For point testing and pulse expansion functions, the relative backscattering amplitude error can be approximated away from internal resonances as

$$\left| \frac{\Delta S(0)}{S(0)} \right| \simeq 1.5 \, n_\lambda^{-2} \tag{4.31}$$

This estimate is compared to observed numerical results in Figure 4.2. While this result is for backscattering, the convergence rate of the bistatic scattering amplitude is also second order with respect to the mesh density.

From these error estimates, it can be seen that the numerical behavior of the scattering amplitude for the MFIE is quite different from that of the EFIE. With the EFIE, as discussed in Section 3.2, the projection error cancels due to the variational property of the scattering amplitude, leaving only the third-order aliasing error in the numerator of (3.47). For the MFIE with the point/pulse discretization, this cancellation does not occur. In (4.28), the projection error is the dominant contribution, leading to a second-order error, which is the same order as the current solution error. For this discretization, it follows that the method of moments for the MFIE is not superconvergent. The reason for this and remedies for the poorer accuracy will be considered in Section 4.5.

4.2 Nonideal Discretizations

The analysis of Section 4.1 assumes a conformal mesh representation of the scatterer geometry and exact integration of moment matrix elements. We will now study the effect on solution error of nonideal discretizations, including numerical integration and the use of a flat-facet mesh.

4.2.1 Quadrature Error

Since the kernel of the integral part of the MFIE operator is continuous, it can be readily integrated to a high accuracy, and a simple midpoint Euler quadrature rule is adequate for matrix evaluation. For low-order basis functions, we will see that only a small number of quadrature points are needed for the implementation to be considered ideal.

To analyze the effect of numerical quadrature, we observe that the integrals in the definition of the basis function Fourier coefficients T_q and F_q in (3.24) and (3.25) arise from the source and observation integrals in the moment matrix element (2.28). We can account for the testing and expansion quadrature rules by replacing the integrals with the quadrature rule. For the point/pulse discretization, this yields the periodic sinc function (3.59). For the projection error, the periodic sinc function need only be evaluated for small values of q, so the periodic sinc function may be approximated by the principle period as

$$F_{q,M} \simeq F_q \left[1 + \frac{1}{3} \left(\frac{\pi q}{M k_0 a n_\lambda} \right)^2 \right] \qquad (4.32)$$

Substituting this into the definition of the projection error (4.10), we see that to leading order

$$E_{q,M}^{(1)} \simeq E_q^{(1)} + \frac{1}{3} \left(\frac{\pi q}{M k_0 a n_\lambda} \right)^2 \qquad (4.33)$$

Comparing (4.33) with the projection error (4.22) for the ideal case shows that quadrature error for the MFIE only augments the second-order term of the projection error, whereas quadrature introduces a dominant first-order contribution for the EFIE.

By inserting (4.32) into (4.11) and evaluating the summation over s, the aliasing error contribution can be shown to be

$$E_{q,M}^{(2)} \simeq (-1)^{M+1} \frac{(k_0 a)^2}{2\lambda_q^{\mathcal{M}_{TM}}} F_{q,M} (Mk_0 a n_\lambda)^{-3} \tag{4.34}$$

Though aliasing error has increased from fourth to third order in n_λ, this error term is still negligible compared to the projection error (4.33). From this analysis, it can be seen that the spectral error for the MFIE remains second order when moment matrix elements are integrated using a low-order quadrature rule. It follows from this analysis that quadrature error does not change the order of convergence of the current solution.

For the scattering amplitude, we must consider the effect of the quadrature rule on the scattered plane wave, which is discretized using the expansion functions f_n. We can account for this by replacing the integral in (2.46) with the quadrature rule. In the end, however, since the dominant error contribution is the second-order projection error in (4.30), the scattering amplitude is not significantly affected.

For both the current and scattering amplitude solutions, quadrature error for the MFIE does not significantly increase the solution error. This is in contrast to the EFIE, for which the solution error for both current and scattering amplitude solutions worsened to first order. The reason for this is the smoothness of the kernel of the integral operator $\mathcal{M}_{TM}$ in the MFIE. With such a smooth kernel, integrals can be evaluated accurately with a simple, low-order quadrature rule.

4.2.2 Single Integration Point

For a single integration point, the above analysis breaks down. Interestingly, the solution accuracy in this special case actually becomes better by orders of magnitude than is obtained with exact integration of moment matrix elements. For one integration point ($M = 1$), $F_q T_{-q} = 1$ in (4.10) and the projection error vanishes. Since the aliasing error associated with the very smooth kernel of the integral part of the MFIE operator is exceptionally small (third order in n_λ^{-1}), the current and scattering solution errors converge very rapidly.

This rapid convergence presumes that the integral term of the MFIE is evaluated using the single-point integration rule for diagonal moment matrix elements as well as off-diagonal elements. Evaluating the integral term with the single-point integration rule results in the diagonal moment matrix element

$$Z_{mm} = \frac{1}{2} - \frac{h}{4\pi} \kappa(\rho_m) \tag{4.35}$$

where $\kappa(\rho)$ is the curvature of the scatterer at ρ. The sign of κ is positive where the scatterer surface is concave and negative where it is convex. The term $-\kappa h/(4\pi)$ is commonly referred to as a curvature correction (e.g., [3]). It is common in practice to take the diagonal elements to be identically 1/2 and to neglect the contribution from the integral operator, but if this is done, the order of accuracy becomes poorer and the unusual third-order convergence is not obtained. The curvature correction must be included in the diagonal moment matrix elements in order to obtain third-order solution convergence [1].

4.2.3 Geometrical Discretization Error

We will now relax the assumption of a conformal geometrical model for the scatterer and consider error caused by the use of a flat-facet mesh. For a circular cylinder and a point/pulse discretization, with an exact geometrical model the diagonal element (4.35) becomes

$$Z_{mm} = \frac{1}{2} - \frac{1}{2k_0 a n_\lambda} \tag{4.36}$$

since the curvature of the cylinder is equal to the radius a. For a flat-facet mesh, the integral part of the MFIE operator is zero, since $\cos \psi = 0$ in (2.19), and the diagonal elements moment matrix elements are identically 1/2. This leads to a shift away from the more accurate diagonal element (4.35). Since a shift in the diagonal elements of a matrix leads to a like shift in the spectrum, the spectral error for a flat-facet mesh is

$$E_q \simeq -\frac{1}{2k_0 a n_\lambda \lambda_q^{\mathcal{M}_{\text{TM}}}} \tag{4.37}$$

which is a first-order error contribution. The first-order spectral error translates into first-order current and scattering amplitude errors.

By making use of the geometrical discretization error (4.37) in (4.23), we can determine the effect of a flat-facet mesh on the current solution error. Assuming that the frequency is not near an internal resonance of the scatterer, we find that

$$\frac{\|\Delta J\|_{RMS}}{\|J\|_{RMS}} \simeq 1.5 \, (k_0 a)^{-1} n_\lambda^{-1} \tag{4.38}$$

For the backscattering amplitude error,

$$\left| \frac{\Delta S(0)}{S(0)} \right| \simeq 2.0 \, (k_0 a)^{-1} n_\lambda^{-1} \tag{4.39}$$

As might be expected, these error estimates decrease with $k_0 a$, because the scatterer curvature becomes smaller and the deviation of the flat-facet approximation from the exact geometry becomes less severe as the radius of the scatterer increases. These estimates are compared to numerical results in Figure 4.2.

4.3 CFIE

As observed in Section 2.3, the EFIE and MFIE suffer from increased solution error at frequencies near the internal resonances of the scatterer geometry. For the MFIE, this is evident in Figure 4.1. To avoid this problem, the electric and magnetic field formulations can be combined into an integral equation that does not have any real internal resonance frequencies. To determine solution error for the combined field formulation, we can make use of the error analyses for the EFIE and MFIE of Chapter 3 and Section 4.1.

The key to the CFIE error analysis is that for a circular cylinder, the EFIE and MFIE share the same eigenfunctions. Because of this, the eigenvalues of the CFIE operator are given by a weighted combination of EFIE and MFIE eigenvalues according to

$$\lambda_q^{\mathcal{C}} = \frac{\eta\pi ka}{2} H_q(ka)[\alpha J_q(ka) - j(1-\alpha)J_q'(ka)] \tag{4.40}$$

where α is the CFIE weighting constant in (2.22) chosen on the interval $[0,1]$. The same linear combination holds for the eigenvalues of the moment matrix.

Based on (4.40), it is straightforward to show that the relative spectral error for the CFIE is

$$E_q^{\mathcal{C}} = \frac{\alpha J_q(ka)E_q^{\mathcal{L}} - j(1-\alpha)J_q'(ka)E_q^{\mathcal{M}}}{\alpha J_q(ka) - j(1-\alpha)J_q'(ka)} \tag{4.41}$$

in terms of the EFIE and MFIE relative spectral errors $E_q^{\mathcal{L}}$ and $E_q^{\mathcal{M}}$ obtained above and in Chapter 3. The current and scattering amplitude error expressions (4.21) and (4.29) are also valid for the CFIE by substituting in the spectral error (4.41). From (4.41), it is apparent that the spectral error as well as the current and scattering amplitude errors will inherit the lower of the convergence rates of the EFIE and MFIE.

By fixing $\alpha = 0.2$ and assuming a point/pulse discretization, we can use (4.41) to compute simple error estimates for the CFIE. For an ideal discretization with no quadrature error and a conformal geometrical model, the current and scattering amplitude solution errors are

$$\frac{\|\Delta J\|_{RMS}}{\|J\|_{RMS}} \simeq 0.3\, n_\lambda^{-2} \tag{4.42}$$

$$\left|\frac{\Delta S(0)}{S(0)}\right| \simeq 0.3\, n_\lambda^{-2} \tag{4.43}$$

It can be seen from these estimates that the error in both cases is second order.

For a nonideal discretization, the use of a low-order, M-point quadrature rule limits the CFIE solution convergence rate to the first-order convergence of the EFIE:

$$\frac{\|\Delta J\|_{RMS}}{\|J\|_{RMS}} \simeq 0.3 \, (Mn_\lambda)^{-1} \tag{4.44}$$

$$\left| \frac{\Delta S(0)}{S(0)} \right| \simeq 0.4 \, (Mn_\lambda)^{-1} \tag{4.45}$$

For a flat-facet mesh, error is limited by the sensitivity of the MFIE to the curvature of the scatterer. The solution error estimates are

$$\frac{\|\Delta J\|_{RMS}}{\|J\|_{RMS}} \simeq 1.2 \, (ka)^{-1} n_\lambda^{-1} \tag{4.46}$$

$$\left| \frac{\Delta S(0)}{S(0)} \right| \simeq 1.5 \, (ka)^{-1} n_\lambda^{-1} \tag{4.47}$$

We can see that the price paid for the elimination of error at internal resonance frequencies is that solution error degrades to the poorer of the solution errors obtained with the EFIE or MFIE alone.

These error estimates assume that the same discretization scheme is used for both the MFIE and EFIE components of the CFIE. It is possible to decrease error by mixing discretization schemes and choosing the basis set that is best suited to the properties of each integral operator. By discretizing the MFIE with a single-point integration rule augmented by a curvature correction term for diagonal elements of the moment matrix, and the EFIE with exact numerical integration of moment matrix elements, a third-order convergence rate for the CFIE scattering amplitude solution can be achieved. The solution convergence rates can also be improved by making use of higher-order basis functions.

4.4 SOLUTION ERROR FOR OTHER SMOOTH SCATTERERS

For the EFIE, it could be demonstrated numerically that the solution error estimates developed for the circular cylinder extend to other scatterer geometries. As might be expected, the same observation holds for the MFIE and CFIE. Figures 4.3 and 4.4 show error curves for the scatterers depicted in Figure 3.7. Reference solutions are generated using a highly refined mesh, and used to generate the solution error results for coarser meshes shown in the figures.

For all the scatterer geometries, solution convergence rates are the same as those obtained analytically for the circular cylinder. The absolute error for the noncircular geometries, however, can be significantly larger. The sensitivity to geometry is strongest

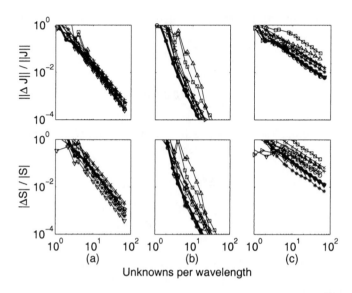

Figure 4.3: Relative current and backscattering amplitude errors for the MFIE with TM polarization for the scatterers shown in Figure 3.7. (a) Ideal discretization. (b) Single-point quadrature rule with curvature correction for moment matrix integrations. (c) Flat-facet mesh. Convergence rates are second, third, and first order, for (a), (b), and (c), respectively. The single-point quadrature rule in (b) is a special case discretization with unusually high accuracy. (©2005 IEEE [1].)

for the flat-facet mesh results in 4.3(c) and 4.4(c). This indicates that accuracy suffers when highly curved scatterers are represented with a flat-facet geometrical approximation.

4.5 SUPERCONVERGENCE AND REGULARIZATION

Chapter 3 explained current and scattering accuracy in terms of projection and aliasing errors. Projection error is associated with the low-order modeled modes of the scattering problem that are well represented by the basis function expansion. High-order, unmodeled modes that oscillate too rapidly to be modeled by the basis functions cause aliasing error. The purpose here is to use these concepts to analyze in greater detail the difference in the effect of projection error on MFIE accuracy relative to that observed with the EFIE.

In the numerical scattering amplitude solution, projection error cancels for the TE-EFIE and TM-EFIE, leaving only the error caused by aliasing of high-order eigenmodes. Consequently, the rate of decay or growth of the high-order eigenvalues either improves or degrades the scattering amplitude error. Since high-order eigenvalues are

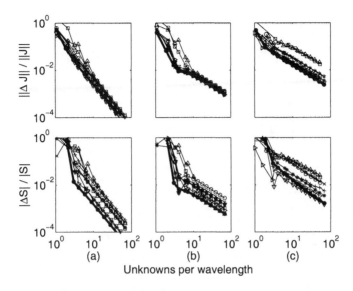

Figure 4.4: CFIE solution errors for the discretizations of Figure 4.3. Convergence rates are second, first, and first order, respectively. Errors are given by the worst case of either the EFIE and MFIE. (©2005 IEEE [1].)

associated with the kernel singularity, aliasing error is determined by the degree of singularity of the kernel. As observed in Section 3.2, the cancellation of the projection error is a manifestation of the variationality of the method of moments for the numerical scattering amplitude computation. For the TM-MFIE and TE-MFIE with low-order basis functions, the projection error does not cancel. The accuracy of the scattering amplitude is much poorer than obtained with the EFIE, and is no better than that obtained for the surface current solution. In this case, the numerical scattering amplitude is not superconvergent.

After developing a deeper insight into the reason for the lack of cancellation of the projection error for the MFIE, we will use that understanding to develop a simple modification of the MFIE that overcomes this problem and improves the solution convergence rate by as many as three orders [4].

4.5.1 Convergence Rates for EFIE and MFIE

To provide a context for understanding the poor accuracy of the discretized MFIE for low-order basis functions, it is helpful to consider a sequence of discretizations with increasing polynomial order for the TM-EFIE, TE-EFIE, and MFIE. Scattering amplitude convergence rates for several different choices of testing and expansion functions with ideal discretizations (exact integration of moment matrix elements and a conformal mesh) are given in Table 4.1. The testing and expansion functions are indicated pictorially, along with their combined polynomial order in the first two rows of the table.

Table 4.1: Scattering amplitude solution convergence rates for smooth scatterers with ideal moment method discretizations [1, 5]. The pair of symbols in the column headings indicate the testing function and expansion functions, and h is the mesh discretization width. The shaded entries are irregular cases as discussed in the text.

Basis functions (↑ delta, ⊓ pulse, ∧ hat)	↑ ⊓	⊓⊓ or ↑ ∧	⊓∧ or ∧⊓	∧ ∧
Combined polynomial order $p + p'$	−1	0	1	2
TM-EFIE scattering error	h^3	h^3	h^4	h^5
TM/TE MFIE scattering error	h^2	h^2	h^4	h^4
TE-EFIE scattering error	h^3	h^3	h^3	h^3

4.5.1.1 Regular Cases

For some of the discretizations in Table 4.1, particularly for higher-order expansion and testing functions, the observed solution convergence rates can be explained using the aliasing error defined in (3.18) as

$$E_q^{(2)} = \frac{1}{\lambda_q} \sum_{s \neq 0} \lambda_{q+sN} T_{-q-sN} F_{q+sN} \tag{4.48}$$

This expression holds for all of the integral equations, with the substitution of the appropriate operator eigenvalues λ_q. Except for the two lowest-order MFIE cases, the projection error part of the spectral error cancels and the scattering amplitude error is determined solely by the aliasing error. For the unshaded boxes in the table, the series in (4.48) is rapidly convergent and can be approximated by the $s = 1$ term. In other words, for these discretizations the first aliased mode is dominant and the others can be neglected. Thus, we can predict solution convergence rates using simple order arguments based on asymptotic approximation of the operator eigenvalue and Fourier coefficients of the basis functions.

For the TM-EFIE with a discretization of order $p + p' = 0$, the orders of the factors with respect to h in the $s = 1$ term of the spectral error are

$$\lambda_{q+N}^{\mathcal{L}} \simeq \frac{j\eta\pi k_0 a}{2N} \sim h$$

$$T_{q+N} = \frac{\sin[\pi(1+q/N)]}{\pi(1+q/N)} \simeq -\frac{q}{N} \sim h$$

$$F_{q+N} \sim h$$

since $N = 2\pi a/h$. We have considered the case of pulse testing and expansion functions. For point testing with triangle expansion functions, $T_q = 1$ and $F_q \simeq 1/q^2$, so the product of the orders and hence the overall solution convergence rate are the same. From these order expressions, it can be seen that the third-order convergence of this discretization is associated with the $1/q$ asymptotic falloff of the eigenvalue λ_q and the $1/q$ falloff of the Fourier coefficients of the testing and expansion functions.

For the MFIE with a discretization of order $p + p' = 2$, the orders of the factors with respect to h in the $s = 1$ term of the spectral error are

$$\lambda_{q+N}^{\mathcal{M}} \simeq \frac{1}{2}$$

$$T_{q+N} = \left\{ \frac{\sin[\pi(1+q/N)]}{\pi(1+q/N)} \right\}^2 \sim h^2$$

$$F_{q+N} \sim h^2$$

These results hold for both the TE and TM polarizations. Combining the factors leads to the error order estimate h^4, which matches that given in Table 4.1 as well. For the TE-EFIE and a discretization of order $p + p' = 2$, we have

$$\lambda_{q+N}^{\mathcal{N}} \sim h^{-1}$$

$$T_{q+N} \sim h^2$$

$$F_{q+N} \sim h^2$$

which leads to the error order estimate h^3.

As discussed in Section 4.1.1, the decay rate of the operator eigenvalues is associated with the smoothness of the kernel, and the falloff of the expansion and testing Fourier coefficients is determined by the smoothness of the basis functions. These considerations indicate that for the regular (unshaded) cases in Table 4.1, scattering amplitude error is determined by the combined smoothness of the operator kernel and the testing and expansion functions. The smoother the kernel and basis functions, the more accurate the numerical solution, and the solution convergence rate is the sum of the decay rates of the Fourier representations of the kernel and basis.

In view of this, considering the TM-EFIE with an order $p + p' = 0$ discretization as an example, the convergence order argument can be applied even more simply by inspecting the moment matrix element

$$Z_{mn} = \frac{1}{h} \int ds \underbrace{t_m(s)}_{O(k^{-1})} \underbrace{\mathcal{L}}_{O(k^{-1})} \underbrace{f_n}_{O(k^{-1})} \qquad (4.49)$$

and considering the indicated asymptotic decay rates of the Fourier representations of each factor, for pulse expansion and testing functions. The Fourier transform of a pulse function is the sinc function given in (2.58), and decays at a k^{-1} rate for large spatial frequencies. The $s = 1$ term of the aliasing error (4.48) is essentially equivalent to evaluating the Fourier transforms at twice the mesh Nyquist frequency k_{max} given by (2.63). Since $k_{max} \sim 1/h$, the $O(k^{-3})$ total decay rate of the Fourier transforms in (4.49) translates to an aliasing error of order h^3, the same result obtained above (the leading factor of h^{-1} in (4.49) combines with the integration over ds and does not contribute to the convergence order argument). These arguments lead to a fundamental principle for MoM error behavior: the convergence rate of the scattering amplitude for a smooth scatterer with an ideal implementation of the moment method is determined by the asymptotic decay rate of the product of the Fourier transforms of the expansion and testing functions and the operator kernel.

Finally, we observe that for the regular cases, the improvement in solution accuracy with increasing kernel smoothness has been observed in the literature. The increase in solution error in the progression TM-EFIE $\rightarrow$ MFIE $\rightarrow$ TE-EFIE was observed for all discretizations except the $p = 0$ case in [6], despite the differences between the discretization schemes considered in that reference and those treated here. For 3D scattering problems with low-order basis functions, a similar comparison with respect to MFIE and EFIE scattering convergence rates has been reported [7,8].

4.5.1.2 Irregular Cases

For the irregular (shaded) cases in Table 4.1, the order argument used above breaks down. For the EFIE, the signs of the terms in the series (4.48) alternate in such a way that the sum is smaller than might be expected. In some of the irregular cases, the series is not absolutely convergent, which can make the discretization unstable [9]. As discussed in Section 3.4.2, pulse functions ostensibly do not provide sufficient continuity to evaluate the derivatives in the second term of (2.11). Due to a cancellation of the field radiated by the delta function doublet charge distribution associated with a pulse current, however, convergent results can be obtained. These discretizations may not give meaningful results for irregular meshes.

For the two lowest-order MFIE discretizations in Table 4.1, the irregularity of the solution convergence rate has a different cause—failure of the superconvergence property—which will be discussed in the next section.

4.5.2 Nonsuperconvergent Cases

For the two lowest-order MFIE cases with $p + p' = -1$ and $p + p' = 0$, the scattering amplitude convergence rate is second order. Since this convergence rate is the same as that of the current solution error, it follows that for the point/pulse and pulse/pulse discretizations, the moment method implementation is not superconvergent. In Section 4.1.7, the failure of the superconvergence property is linked to the identity operator in the MFIE and the lack of cancellation of the projection error in (4.30) for low-order, orthogonal basis functions.

We can provide another perspective on the failure of superconvergence for the MFIE with low-order basis functions using the variational property of the method of moments. The first question is whether or not a variational principle exists for the MFIE. While this topic has received very little attention in the literature, a functional and transposed operator for the MFIE relative to the symmetric product have been obtained [10]. The connection between the functional and the scattering amplitude is clearest if the MFIE operator is modified to include a cross product with the scatterer surface normal, so that (2.16) becomes

$$\mathcal{L}_m \mathbf{J}_s = \mathbf{H}_t^{\text{inc}} \tag{4.50}$$

where $\mathbf{H}_t^{\text{inc}}$ is the tangential part of the incident magnetic field on the scatterer surface. The adjoint problem is

$$\mathcal{L}_m^a \mathbf{M}_s = \mathbf{E}_t^s \tag{4.51}$$

where we have identified the unknown source as a magnetic current based on its units. The adjoint operator is defined by

$$\langle \mathbf{f}_1, \mathcal{L}_m \mathbf{f}_2 \rangle = \langle \mathcal{L}_m^a \mathbf{f}_1, \mathbf{f}_2 \rangle \tag{4.52}$$

The functional (3.49) becomes

$$I_m(\mathbf{J}_s, \mathbf{M}_s) = \langle \mathbf{E}^s, \mathbf{J}_s \rangle + \langle \mathbf{M}_s, \mathbf{H}^{\text{inc}} \rangle - \langle \mathbf{M}_s, \mathcal{L}_m \mathbf{J}_s \rangle \tag{4.53}$$

The adjoint operator may be constructed explicitly if desired using the definition (4.52). Following the derivation of (3.56), it can be shown that

$$\Delta S = \langle \Delta \mathbf{M}_s, \mathcal{L}_m \Delta \mathbf{J}_s \rangle \tag{4.54}$$

where we have dropped an unimportant constant factor.

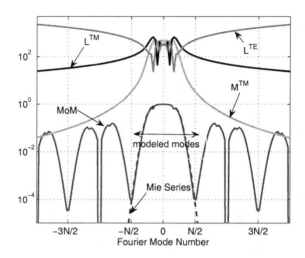

Figure 4.5: Spectrum for the Mie series electric current on a circular cylinder compared to the spectrum of the MoM current solution for pulse expansion functions. The operator eigenvalues are also shown to illustrate the windowing effect of the operator on the current solution.

We will first consider the case of the EFIE, in order to contrast the different behavior observed with the MFIE. For the TM-EFIE, the scattering amplitude error is

$$\Delta S = \langle \Delta J_z^a, \mathcal{L} \Delta J_z \rangle \tag{4.55}$$

As a continuous function, the current error in (4.55) is mostly associated with the expansion functions f_n. For low-order basis functions, the current error is discontinuous, and from a spectral point of view, much of the solution error is contained in high spatial frequencies associated with the discontinuity of the basis. For the TM-EFIE, the integral operator $\mathcal{L}$ is a smoothing operator. In the spatial domain, the smoothing property of the TM-EFIE operator implies that while f_n may be discontinuous, $\mathcal{L}f_n$ is a continuous function of position. As a consequence, the operator $\mathcal{L}$ suppresses the interpolation error in ΔJ, and the variational error formula $\langle \Delta J^a, \mathcal{L} \Delta J \rangle$ is smaller than, say, $\langle \Delta J^a, \Delta J \rangle$. This is represented in Figure 4.5. The operator eigenvalues decay with mode number, which reduces the Fourier domain "sidelobes" of the current solution at mode orders beyond $\pm N/2$. It can be argued that the smoothing property of the operator increases the effective smoothness of the expansion functions when the scattering amplitude is computed.

For the MFIE, the presence of the identity operator in the second-kind integral equation leads to a different conclusion. In Figure 4.5, the operator spectrum of the integral part $\mathcal{M}$ of the MFIE can be seen to decay rapidly with order, but the identity operator makes the eigenvalues of the MFIE approach a constant for large orders. For the 2D TM-MFIE or TE-MFIE operators, (4.54) becomes

$$\Delta S = \tfrac{1}{2}\langle \Delta M, \Delta J \rangle + \langle \Delta M, \mathcal{M}\Delta J \rangle \tag{4.56}$$

Without a smoothing operator in the first term of this expression, the spectral content of ΔJ associated with the discontinuities of the expansion functions is not suppressed, and the scattering amplitude error is larger than for the EFIE. This is reflected in the lowest-order cases for the MFIE in Table 4.1.

Finally, it is interesting to observe that the treatment of [10] motivated a choice of testing functions for the MFIE that effectively zeros out the identity term of the MFIE operator. This is equivalent to discretizing the operator $\mathcal{L}_m$ in (4.50) with identical testing and expansion functions, in which case the cross product with the surface normal included in the operator causes the contribution of the identity term to the moment matrix to vanish. The resulting linear system is ill-conditioned, but it can still be solved in some cases using an iterative algorithm to yield accurate RCS solutions. This further verifies that the identity operator is the cause of the failure of superconvergence for the MFIE.

4.5.3 First- and Second-Kind Operators

Based on these arguments, it is clear that the underlying cause of the poor accuracy for the nonsuperconvergent cases with second-order error in Table 4.1 is the identity operator in the MFIE. With the EFIE, both polarizations, one corresponding to a smoothing operator and the other to an antismoothing operator, lead to cancellation of the projection error and have better accuracy for the scattering amplitude. In spite of this, it is misleading to view the differences between the MFIE and EFIE as merely a consequence of first-kind versus second-kind operators.

From the illustration in Figure 4.5, it would seem that the TE-EFIE would have a much larger scattering amplitude error, since the antismoothing operator with increasing eigenvalues should amplify the sidelobes of the current solution. By the rigorous analysis of Section 3.4, however, there is a cancellation in the sum over high-order modes, and the relatively good accuracy reflected in Table 4.1 for the TE-EFIE is observed in practice. As noted above, this cancellation is unstable and may not occur for irregular meshes. These considerations highlight the subtleties associated with the irregular cases in Table 4.1. Convergence rates can be better (TE-EFIE) or poorer (MFIE) than expected based on simple operator smoothness arguments. If this spectral error cancellation did not occur for the TE-EFIE, however, the larger error suggested by Figure 4.5 would indeed be observed, and error behaviors would be explainable in terms of operator smoothness—not the formal first- or second-kind appearance of the integral equation.

The reason that simple first-kind and second-kind arguments are dangerous here is simply that the TE-EFIE is not a true first-kind integral equation, because the operator $\mathcal{N}$ has a hypersingular kernel. Even the MFIE operator (or any other second-kind operator) can be written in a first-kind form by using the Dirac delta function,

$$\left(\frac{\mathcal{I}}{2} + \mathcal{M}\right) J = \int \left[\frac{1}{2}\delta(\rho - \rho') + M(\rho, \rho')\right] J(\rho')\, d\rho' \qquad (4.57)$$

but because of the singularity of the delta function the operator is not of the first kind in a rigorous sense.

A better classification of the operators is in terms of the asymptotic behavior of the eigenvalues and the associated smoothing properties discussed in Section 4.1.1. More abstract but ultimately equivalent is the Sobolev space classification in terms of the difference in the smoothness classes of the operator domain and range spaces. For the regular cases of Section 4.5.1.1, scattering amplitude error tracks this operator smoothness index in a very simple way. With this as background, we now turn to remedies for the poorer convergence of the MFIE at low orders.

4.5.4 Higher-Order Basis Functions

From Figure 4.5, it can be seen that one approach to reducing the scattering amplitude error (4.56) is to choose basis functions that reduce the high spatial frequency content of the current solution, so that the sidelobes in the figure are suppressed. This occurs for smoother basis functions with a high polynomial order, as reflected by the $p + p' = 1$ and $p + p' = 2$ entries in Table 4.1 for the MFIE.

Returning to the first-kind and second-kind classification, the smoother basis functions can be thought of as regularizing the delta function in the kernel of (4.57). With the kernel effectively smoother, the modified MFIE behaves like a first-kind integral equation and the cancellation of projection error observed for the EFIE occurs for the MFIE. Analytically, the higher-order basis functions change the leading factor in (4.22), causing the term containing the projection error $E_q^{(1)}$ to vanish from (4.30).

While the use of higher-order basis functions to improve accuracy with the MFIE has been demonstrated in practice, we can also develop an alternative approach that increases solution accuracy while retaining the simplicity of low-order basis functions.

4.5.5 High-Order Convergence with Low-Order Basis Functions

For low-order discretizations, the foregoing results point to the identity operator in the MFIE as the culprit for poor accuracy. This is a surprising result, since the identity operator is easily discretized, requires trivial quadrature for integration in moment matrix elements, and discretizes to a well-conditioned matrix. It has also been observed that there is a correlation between operator smoothing properties and solution error, with the more strongly smoothing cases in Section 4.5.1.1 corresponding to higher accuracy. In Section 4.5.4, it is argued that for high-order basis functions, the effective delta function kernel in the MFIE singularity is tempered by the basis functions, and the MFIE effectively becomes a strongly smoothing operator. This leads to a cancellation of eigenfunction projection error in the scattering amplitude and improved solution accuracy.

For low-order basis functions, which do not have a sufficient smoothing effect on the operator to provide high accuracy, one could instead regularize the delta function in (4.57) by replacing it with a generating function, so that $\delta(\rho, \rho') \to D(\rho, \rho')$, where D becomes a delta function as some parameter (in this case, the mesh size h) goes to zero. Under this substitution, the kernel of (4.57) becomes smoother, the MFIE becomes a true first-kind integral equation, and an accurate scattering amplitude solution can be obtained even for low-order discretizations. The MFIE scattering error would then depend on aliasing error only, which is very small due to the $1/q^3$ eigenvalue falloff rate of the strongly smoothing operator $\mathcal{M}$. Since this replacement reduces the effective singularity of the kernel associated with the identity operator in the MFIE, we refer to this approach as regularization of the identity [4, 11].

The key to making the substitution $\delta \to D$ is to ensure that the error introduced by this substitution does not outweigh the benefits of projection error cancellation. To derive a form for the smoothed delta function, it suffices to consider only the identity part of the integral equation, which is

$$\frac{1}{2}J = H_t^{\text{inc}} \tag{4.58}$$

We assume that D has a Fourier series representation of the form

$$D(\rho, \rho') = \sum_q D_q e^{-jq(t-t')} \tag{4.59}$$

where t is a parameter for a closed scatterer surface such that $\rho(t)$ maps the interval $0 \le t \le 2\pi$ to the scatterer contour. Following the manipulations of this and previous chapters, the eigenvalues of the moment matrix generated by discretizing (4.58) are given by

$$\hat{\Lambda}_q = \sum_s D_{q+sN} T_{-q-sN} F_{q+sN} \tag{4.60}$$

The approximate current and scattering amplitude solutions for the identity equation can then be computed as before, yielding the "scattering amplitude" error

$$\Delta S = \sum_q \sum_s (-1)^s A_q^* B_{q+sN} \left[\delta_{s0} - \frac{F_q T_{-q-sN}}{\hat{\Lambda}_{q+sN}} \right] \tag{4.61}$$

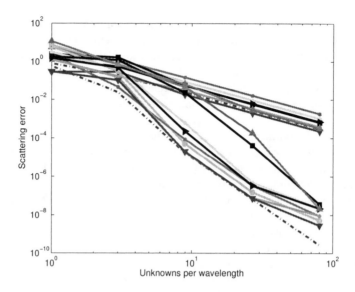

Figure 4.6: TM-MFIE MoM scattering amplitude error for nine smooth scatterers in Figure 3.7. The error curves that decay as h^2 were generated from the unregularized MFIE. The error with the regularized MFIE decay as h^5. TE results are similar. Regularization leads to a three-order-of-magnitude improvement in accuracy. (©2005 IEEE [4].)

where A_q and B_q are the Fourier coefficients of the incident and scattered plane wave fields.

The leading-order behavior of ΔS is determined by the $s = 0$ terms of (4.61), which include the factor

$$1 - \frac{F_q T_{-q}}{\hat{\Lambda}_q} = \left(D_q - 1\right)\frac{F_q T_{-q}}{\hat{\Lambda}_q} + \sum_{s \neq 0} D_{q+sN}\frac{F_{q+sN}T_{-q-sN}}{\hat{\Lambda}_{q+sN}} \qquad (4.62)$$

We can force (4.62) to zero by taking $D_q = 1$ for the N lowest-order modes and 0 otherwise. This gives a very small ΔS, which now comprises only the high-order terms ($s \neq 0$) in (4.61). Evaluating (4.59) gives an explicit expression for the regularized identity kernel,

$$D(\rho, \rho') = \frac{1}{C}\frac{\sin\frac{\pi}{h}(t - t')}{\sin\frac{\pi}{C}(t - t')}e^{-j\frac{\pi}{C}(t - t')} \qquad (4.63)$$

where C is the circumference of the cylinder. This is a periodic sinc function with height $1/h$ and main lobe width $2h$, centered at $\rho = \rho'$.

The effect of regularization is to filter out the high frequency content in the delta function associated with the identity operator. This increase in operator smoothness compensates for the lack of smoothness of low-order basis functions. In the scattering amplitude computation, this filtering suppresses the error in the high spatial frequency components of the current solution. Returning to Figure 4.5, the current solution still has sidelobes due to the discontinuous expansion functions, but the operator now has a windowed spectrum that suppresses the sidelobes in the scattering amplitude error (4.54).

Figure 4.6 shows numerical solution error for the point/pulse discretization and a 32-point, Euler quadrature rule. The scattering bodies considered are the same shown in Figure 3.7. Both regularized and unregularized identity operators were used. The unregularized solutions converge as h^2, whereas the regularized solutions converge as h^5. Results are similar for the TE polarization. These results demonstrate that regularization leads to an improvement in the solution accuracy by three orders of magnitude, without requiring the use of high-order basis functions. The regularization method can also be applied to the EFIE [12] as well as three-dimensional scattering problems [11].

4.6 SUMMARY

If the MFIE is discretized using low-order basis functions, both the current and scattering amplitude solutions converge at a second-order rate. This is in contrast to the EFIE, for which third-order convergence of the scattering amplitude can be realized using accurate integration of moment matrix elements and an exact geometrical model for the scatterer. We have linked the poorer convergence of the MFIE to the identity operator. Even though the identity operator is trivial to discretize for orthogonal basis functions, in its effect on solution error the identity operator behaves as if it were an integral operator with a singular kernel. The scattering amplitude solution convergence rate can be improved by using higher-order basis functions. To obtain increased accuracy for the MFIE with low-order basis functions, regularization of the identity operator can be employed.

We have also seen that solution error for the CFIE is related in a simple way to the accuracy of the EFIE and MFIE. The CFIE solution is dominated by the larger of the EFIE and MFIE errors. With regularization, MFIE solution accuracy can be improved to better than that of the EFIE, allowing accurate CFIE results to be obtained using the same low-order basis functions for both the electric and magnetic field parts of the combined formulation.

Perhaps of greatest significance, we have shown that for discretizations that are regular in the sense of Section 4.5.1.1, the scattering amplitude error for a smooth scatterer with an ideal discretization can be predicted with a simple convergence order argument based on the asymptotic decay rates of the Fourier transforms of the operator kernel and basis functions. We will return to this for 3D scattering problems in

Chapter 7. Despite the mathematical complexity of the spectral error analysis for 3D problems, the simple convergence order argument developed here for 2D problems can be readily applied to understand and predict solution errors to the more general case of vector integral equations in three-dimensional space.

REFERENCES

[1] C. P. Davis and K. F. Warnick, "Error analysis of 2D MoM for MFIE/EFIE/CFIE based on the circular cylinder," *IEEE Trans. Ant. Propag.*, vol. 53, pp. 321-331, Jan. 2005.

[2] S. Amini and K. Chen, "Conjugate gradient method for second kind integral equations—applications to the exterior acoustic problem," *Engr. Anal. with Boundary Elements*, vol. 6, pp. 72-77, 1989.

[3] J. V. Toporkov, R. T. Marchand, and G. S. Brown, "On the discretization of the integral equation describing scattering by rough conducting surfaces," *IEEE Trans. Ant. Propag.*, vol. 46, Jan. 1998.

[4] C. P. Davis and K. F. Warnick, "High order convergence with a low-order discretization of the 2-D MFIE," *IEEE Ant. Wireless Propag. Lett.*, vol. 3, pp. 355-358, Dec. 2005.

[5] K. F. Warnick and W. C. Chew, "Accuracy of the method of moments for scattering by a cylinder," *IEEE Trans. Micr. Th. Tech.*, vol. 48, pp. 1652-1660, Oct. 2000.

[6] A. F. Peterson and M. M. Bibby, "Error trends in higher-order discretizations of the EFIE and MFIE," *Proceedings of the IEEE Antennas and Propagation Society International Symposium*, vol. 3A, Washington DC, pp. 52-55, July 2005.

[7] O. Ergul and L. Gurel, "Investigation of the inaccuracy of the MFIE discretized with the RWG basis functions," *Proceedings of the IEEE Antennas and Propagation Society International Symposium*, vol. 3, Monterey, CA, pp. 3393-3396, June 20-25, 2004.

[8] O. Ergul and L. Gurel, "Improving the accuracy of the MFIE with the choice of basis functions," *Proceedings of the IEEE Antennas and Propagation Society International Symposium*, vol. 3, Monterey, CA, pp. 3389-3392, June 20-25, 2004.

[9] M. I. Aksun and R. Mittra, "Choices of expansion and testing functions for the method of moments applied to a class of electromagnetic problems," *IEEE Trans. Micr. Th. Tech.*, vol. 41, pp. 503-509, Mar. 1993.

[10] S. Yan and Z. Nie, "On the Rayleigh-Ritz scheme of 3D MFIE and its normal solution," *Proceedings of the IEEE Antennas and Propagation Society International Symposium*, San Diego, CA, July 2008.

[11] K. F. Warnick and A. F. Peterson, "3D MFIE accuracy improvement using regularization," *Proceedings of the IEEE Antennas and Propagation Society International Symposium*, Honolulu, HI, pp. 4857-4860, June 2007.

[12] K. F. Warnick, G. Kang, and W. C. Chew, "Regulated kernel for the electric field integral equation," *Proceedings of the IEEE Antennas and Propagation Society International Symposium*, vol. 4, Salt Lake City, UT, pp. 2310-2313, July 2000.

Chapter 5

Geometrical Singularities and the Flat Strip

In previous chapters, we have restricted attention to smooth scatterers. It is well known that geometrical singularities such as edges and corners can induce a singular behavior for the surface current that is not well modeled by polynomial basis functions and reduces solution accuracy considerably. In this chapter, we will treat the problem of edge current singularities using the flat conducting strip as a canonical scatterer geometry. We will then consider wedge geometries with the flat strip being an extreme limiting case.

In order to develop a complete error estimate for the flat strip, we will first consider error on the "interior" part of the strip away from edges, and then we will add the error contribution due to current singularities at the edges. Along the way, while treating the interior part of the strip, some results not discussed in previous chapters, such as a magic "1/3" discretization, will be presented. Unlike the case of smooth scatterers, for which the solution convergence rate is strongly dependent on the integral equation formulation, error due to current singularities is determined primarily by the approximation of the singular current using discrete basis functions. We will find that edge and corner errors lead to a dominant error contribution with a slower asymptotic falloff with respect to mesh density than error due to smooth portions of the geometry.

5.1 FLAT STRIP INTERIOR ERROR, TM-EFIE

For a flat conducting strip, the EFIE operator is nonnormal, which means that the eigenfunctions do not form an orthogonal basis for the domain of the operator. Physically, this occurs because Fourier-type oscillatory modes on the scatterer are coupled by edge diffraction and the associated current singularities at the scatterer ends. As with a nonnormal matrix, an orthogonal basis can be constructed from the operator eigenfunctions together with a finite number of adjoint functions for each eigenvalue [1], but the operator does not have a simple spectral decomposition analogous to (4.2) for the circular cylinder.

The nonnormality of the EFIE operator prevents direct application of the modal analysis of previous chapters to the flat strip. One approach to overcoming this difficulty is the use of the static decomposition (1.3), which amounts to approximating the integral operator by its static limit and determining the solution error from the simpler case of the self-adjoint integral operator relating surface charges to electrostatic fields. The limitations of this approach are discussed in Chapter 1. For high frequencies or scatterers with large electrical size, the static limit fails to capture all the relevant physics, and consequently the resulting moment method solution error estimates are inadequate.

To improve on the self-adjoint operator decomposition (1.3), we will employ a decomposition of the form of (1.4), in which $\mathcal{H}$ is a normal operator and $\mathcal{R}$ is a nonnormal perturbation. In this work, the operator $\mathcal{H}$ corresponds to the high-frequency limit of the EFIE operator $\mathcal{L}$ and is closely related to the physical optics approximation. We will show that $\mathcal{R}$ is small enough that the eigenvalues of the normal part $\mathcal{H}$ provide spectral estimates for the integral operator $\mathcal{L}$ and its matrix discretization using the method of moments. This allows the spectral error concepts developed in Chapters 3 and 4 to be applied for the flat strip by approximating the nonnormal operator $\mathcal{L}$ with the normal operator $\mathcal{H}$.

Although it captures more information about error behaviors than the static limit, the normal operator approximation still has its limits. Physical information about edge diffraction is contained in the operator $\mathcal{R}$, so solution error estimates based on $\mathcal{H}$ do not reflect error associated with current singularities at the scatterer edges. Error due to edge singularities must be treated separately and added to the solution error for the interior of the strip away from the scatterer edges.

5.1.1 Normal Operator Approximation

To develop a normal approximation for the operator $\mathcal{L}$, we will work with its infinite-dimensional Fourier representation, for which the matrix elements are given by

$$L_{qr} = d^{-1}\langle e^{-j\beta_q k_0 x}, \mathcal{L} e^{-j\beta_r k_0 x}\rangle \tag{5.1}$$

where d is the width of the strip and $\langle \cdot, \cdot \rangle$ is the L^2 inner product. The normalized spatial frequency β_q is defined by $\beta_q = q/D$, $q = 0, \pm 1, \pm 2, \ldots$, where $D = d/\lambda$ is the size of the strip in wavelengths. Unlike the circular cylinder, the Fourier representation is not diagonal, because the functions $e^{-j\beta_q k_0 x}$ are not eigenfunctions of the operator. The Fourier modes are approximate eigenfunctions, however, and L_{qr} is strongly diagonal. We will take the normal operator $\mathcal{H}$ to be the diagonal elements of L_{qr}, so closed form approximations for the diagonal elements are needed.

The kernel of the integral operator (2.8) can be represented in terms of a 1D Fourier transform as

$$g(x, x') = -\frac{j}{4\pi} \int_{-\infty}^{\infty} \frac{dk_x}{\sqrt{k_0^2 - k_x^2}} e^{-jk_x(x-x')} \qquad (5.2)$$

Inserting this in (5.1) and evaluating the integrals over the scatterer leads to

$$L_{qr} = \frac{\eta}{2\pi^2 D} \int_{-\infty}^{\infty} \frac{d\beta}{\sqrt{1-\beta^2}} \frac{\sin\left[\pi D(\beta - \beta_q)\right]}{\beta - \beta_q} \frac{\sin\left[\pi D(\beta - \beta_r)\right]}{\beta - \beta_r} \qquad (5.3)$$

where $\beta = k_x/k_0$.

We now wish to estimate the integral in (5.3) asymptotically as $D \to \infty$. For $q = r$, (5.3) becomes

$$L_{qq} = \frac{\eta}{2\pi^2 D} \int_{-\infty}^{\infty} \frac{d\beta}{\sqrt{1-\beta^2}} \left\{ \frac{\sin\left[\pi D(\beta - \beta_q)\right]}{\beta - \beta_q} \right\}^2 \qquad (5.4)$$

For large D, the squared sinc function in the integrand is narrow enough that it approaches a delta function, and we can make the approximation

$$L_{qq} \simeq \frac{\eta}{2\pi^2 D \sqrt{1-\beta_q^2}} \int_{-\infty}^{\infty} d\beta \left\{ \frac{\sin\left[\pi D(\beta - \beta_q)\right]}{\beta - \beta_q} \right\}^2$$

$$= \frac{\eta}{2\sqrt{1-\beta_q^2}} \qquad (5.5)$$

More rigorously, in [2], the asymptotic expansion

$$L_{qq} \sim \frac{\eta}{2\sqrt{1-\beta_q^2}} - \frac{\eta}{2\pi^2(1-\beta_q^2)} \left[-j + \frac{\beta_q \ln\left(-j\beta_q + \sqrt{1-\beta_q^2}\right)}{\sqrt{1-\beta_q^2}} \right] D^{-1}$$

$$+ O(D^{-3/2}), \quad D \to \infty \qquad (5.6)$$

is obtained. The first term is the limiting value for scattering by an infinite conducting plane, and higher-order terms represent the effect of diffraction by the edges of the strip. This expression is related to the physical theory of diffraction (PTD) for scattering by a conducting strip.

Equation (5.6) breaks down if $|\beta_q| = 1$. These values of β_q correspond to surface wave current modes with spatial frequency k_0. The radiated fields for these modes travel parallel to the strip. For $|\beta_q| = 1$, the asymptotic expansion becomes [2]

$$L_{qq} \sim \frac{\sqrt{2}\eta}{3}(1+j)D^{1/2} + \frac{\eta}{8\sqrt{2\pi}}(1-j)D^{-1/2}$$

$$+ \frac{\eta}{6\pi^2}\left[j - \frac{7\sqrt{2}}{4}\right]D^{-1} + O(D^{-3/2}), \quad D \to \infty \qquad (5.7)$$

As will be seen in Chapter 9, the surface wave mode determines the condition number of the moment matrix.

These estimates for the diagonal elements L_{qq} provide the Fourier representation of the approximate operator $\mathcal{H}$ in (1.4). The full matrix L_{qr} is equivalent to the operator $\mathcal{L}$, and $\mathcal{H}$ will be the main diagonal of L_{qr}. Since $\mathcal{H}$ is diagonal, it is by construction a normal operator. It remains to show that nonnormal part of the operator, $\mathcal{R} = \mathcal{L} - \mathcal{H}$, is sufficiently small that the numerical properties of the method of moments for the flat strip can be determined using $\mathcal{H}$ rather than the exact operator $\mathcal{L}$.

The infinite matrix representation L_{qr} is not diagonally dominant in the strict sense, so the Gersgorin circle theorem cannot be applied to bound the eigenvalues near the diagonal elements. The diagonal elements are large, however, and a somewhat weaker theorem does provide a useful result. From the Bauer-Fike theorem [3], we have the relative error bound

$$\min_q \frac{|L_{qq} - \lambda_q|}{|L_{qq}|} \leq \|\mathcal{H}^{-1/2}\mathcal{R}\mathcal{H}^{-1/2}\| \qquad (5.8)$$

where λ_q is an eigenvalue of $\mathcal{L}$. It can be shown [2] that the norm on the right of (5.8) tends to a constant as $D \to \infty$, so that the diagonal elements L_{qq} provide estimates of the eigenvalues of $\mathcal{L}$ with bounded error as D becomes large. Physically, the large diagonal elements are associated with specular scattering from the strip, and the off-diagonal elements represent weaker bistatic scattering.

5.1.2 Discretized Operator Spectrum

We will now use the approximate operator $\mathcal{H}$ to study the accuracy of the method of moments for the flat strip. As with the circular cylinder, we will first estimate the eigenvalues of the moment matrix to determine the spectral error caused by discretization. The spectral error will then determine the current and scattering amplitude solution errors.

We will define a mesh for the scatterer in terms of element center points $x_n = (n - 1/2)h - d/2$, where the index n ranges from 1 to N and the total number of degrees of freedom is $N = d/h$. The expansion and testing functions are of the form $f_n(x) = f(x - x_n)$ and $t_n(x) = t(x - x_n)$, with $f(x)$ and $t(x)$ given by (2.30).

Just as the Fourier modes in (5.1) are not eigenfunctions of $\mathcal{L}$ for the flat strip, the sampled Fourier modes $v_{q,m} = e^{-jk_0\beta_q x_m}$ are not eigenvectors of the moment matrix $\mathbf{Z}$.

This is in contrast to the circular cylinder, for which both the continuous operator and moment matrix are normal and have Fourier functions as eigenfunctions or eigenvectors. Fortunately, in a way that is analogous to (5.8), the sampled plane wave modes are approximate eigenvectors, so they can be used to obtain eigenvalue estimates.

Transforming the moment matrix using sampled plane wave or Fourier modes leads to

$$\hat{L}_{qr} = \frac{1}{N} \sum_{m,n=1}^{N} e^{jk_0\beta_q x_m} Z_{mn} e^{-jk_0\beta_r x_n} \tag{5.9}$$

We can view this as a unitary transformation that approximately diagonalizes the moment matrix. With $\beta_q = q/D$ and $D = d/\lambda$, the vectors $e^{-jk_0\beta_q x_m}$ form an orthogonal basis. Since $\hat{L}_{qr}$ is strongly diagonal, the diagonal elements will provide eigenvalue estimates for the moment matrix.

By making use of (5.2), the moment matrix elements can be written as

$$Z_{mn} = \frac{\eta}{2n_\lambda} \int_{-\infty}^{\infty} \frac{d\beta}{\sqrt{1-\beta^2}} e^{-j2\pi\beta(m-n)/n_\lambda} T(-\beta)F(\beta) \tag{5.10}$$

where $T(\beta)$ and $F(\beta)$ are the Fourier transforms of the testing and expansion functions $t(x)$ and $f(x)$, normalized by $1/h$:

$$T(\beta) = \frac{1}{h} \int e^{jk_0\beta x} t(x)\, dx \tag{5.11}$$

$$F(\beta) = \frac{1}{h} \int e^{jk_0\beta x} f(x)\, dx \tag{5.12}$$

These definitions are essentially identical to (2.58), except that they are given in terms of normalized spatial frequency β rather than k.

The transformed moment matrix representation can then be expressed as

$$\hat{L}_{qr} = \frac{\eta}{2n_\lambda^2 D} \int_{-\infty}^{\infty} \frac{d\beta}{\sqrt{1-\beta^2}} B_q(\beta) B_r(\beta) T(-\beta) F(\beta) \tag{5.13}$$

where

$$B_q(\beta) = \frac{\sin[\pi D(\beta - \beta_q)]}{\sin[\pi(\beta - \beta_q)/n_\lambda]} \tag{5.14}$$

is a periodic sinc function or a scaled Dirichlet function. The task now is to approximate the integral in closed form.

The function $B_q(\beta)$ approaches a periodic series of delta functions as D becomes large. The delta functions are located at the zeros of the denominator of B_q, which occur at $\beta_{q,s} = \beta_q + sn_\lambda$, $s = 0, \pm1, \pm2, \ldots$. The integral in (5.13) can be evaluated

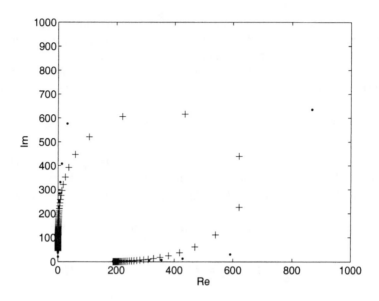

Figure 5.1: Spectrum of the moment matrix for the TM-EFIE for a flat strip, $D = 20$. Basis: point testing, pulse expansion functions. Mesh: flat-facet, $n_\lambda = 10$. Moment matrix integration: exact. Pluses: computed eigenvalues. Dots: eigenvalue estimates given by the normal operator approximation, (5.6) and (5.7).

approximately by expanding the integrand about each of the maxima of B_q, which leads to

$$\hat{L}_{qq} \simeq \sum_s T(-\beta_{q,s})F(\beta_{q,s})L_{q+sN,q+sN}, \quad D \to \infty \tag{5.15}$$

This expression has the same form as (3.13) for the circular cylinder, except that it is given in terms of eigenvalue estimates rather than exact operator eigenvalues.

Based on the same arguments as used in the previous section for the continuous operator, the N eigenvalues of the moment matrix $\mathbf{Z}$ can be estimated by $\hat{\lambda}_q \simeq \hat{L}_{qq}$, $-N/2 + 1 \leq q \leq N/2$, or $-(N-1)/2 \leq q \leq (N-1)/2$ if N is odd. As discussed for the cylinder in Section 3.1, the eigenvalues λ_q of $\mathcal{L}$ with order $|q| > N/2$ are unmodeled eigenvalues, and represent eigenfunctions with spatial frequency content beyond the approximating power of the basis functions. For the unmodeled modes, the normalized spatial frequency is

$$\beta_q > \frac{N}{2D} = \frac{n_\lambda}{2} \tag{5.16}$$

and the spatial frequency $k_0\beta_q$ is greater than the mesh Nyquist frequency (2.63).

Figure 5.1 compares the numerically computed spectrum of the moment matrix for a strip of width 20λ to the approximation $\hat{\lambda}_q \simeq \hat{L}_{qq}$. It can be seen that there is a rather significant difference between the analytical approximation and the computed eigenvalues, but as the goal here is error analysis, rather than an exact representation of the operator, a rough eigenvalue estimate suffices.

It is interesting to consider the physical meaning of different parts of the operator spectrum. The spectral estimates developed above amount to approximating the operator eigenfunctions as modes of the form $e^{-j\beta k_0 x}$, where β is a dimensionless parameter giving the spatial frequency of the mode. For high spatial frequencies ($|\beta| \gg 1$), the eigenvalues are approximately equal to $j(\eta/2)/\sqrt{\beta^2 - 1}$. In Figure 5.1, these eigenvalues approach the origin from the positive imaginary axis. Since these modes radiate evanescent fields, the real part of the eigenvalue is small. For the continuous operator $\mathcal{L}$, there are an infinite number of these imaginary eigenvalues with the origin as an accumulation point. Global geometrical information is contained in the spectrum of the low frequency modes only, since the behavior of the eigenvalues of high frequency modes is independent of geometry, as can be seen by comparing Figures 5.1 and 3.1 for the flat strip and cylinder.

The low-frequency modes ($|\beta| < 1$) have eigenvalues given by $(\eta/2)/\sqrt{1 - \beta^2}$, and approach $\eta/2$ on the real axis for the DC mode ($\beta = 0$). If the width of the strip were infinite, there would be no coupling between the low and high frequency modes, and the eigenvalues would lie entirely on the real and imaginary axes. Due to edge diffraction, the modes are coupled, and the two parts of the spectrum join to form a loop in the complex plane. The surface wave mode ($|\beta| = 1$) has the largest magnitude. Since the eigenvalue is large, the amplitude of this mode in the current solution is suppressed. For the TM-EFIE, the surface wave mode can be considered to be antiresonant. Since the surface wave mode eigenvalue grows in magnitude with the electrical size D, it will be seen in Chapter 9 that this mode causes an increase in moment matrix condition number with electrical size.

The approximate eigenvalues L_{qq} are degenerate for $\pm q$, whereas the eigenvalues λ_q of $\mathcal{L}$ are distinct, since the nonnormal operator perturbation $\mathcal{R}$ removes the degeneracy of the even and odd modes $\cos(k_0\beta_q x)$ and $\sin(k_0\beta_q x)$. For this reason, each eigenvalue estimate (dot) in Figure 5.1 represents a pair of computed eigenvalues (pluses). We will now use these eigenvalue estimates to analyze the spectral error introduced by discretization.

5.1.3 Spectral Error

In the previous section, we developed spectral estimates for a normal approximation to the continuous operator L_{qq} and a normal approximation of the moment matrix $\hat{L}_{qq}$. The eigenvalue shift $\Delta L_{qq} = \hat{L}_{qq} - L_{qq}$ provides an estimate of the spectral error $\Delta\lambda_q = \hat{\lambda}_q - \lambda_q$ introduced by discretization. From (5.15), the approximate relative spectral error $E_q \simeq \Delta L_{qq}/L_{qq}$ is

$$E_q(n_\lambda, \alpha) \simeq T(-\beta_q)F(\beta_q) - 1 + j\sqrt{1-\beta_q^2} \sum_{s\neq 0} \frac{T(-\beta_{q,s})F(\beta_{q,s})}{\sqrt{\beta_{q,s}^2 - 1}} \tag{5.17}$$

where $\beta_{q,s} = \beta_q + sn_\lambda$. In deriving this expression, we have retained only the leading term in the asymptotic expansion of L_{qq}.

Following the terminology introduced in Section 3.1.3, the first term, $T(-\beta_q)F(\beta_q) - 1$, is projection error due to inaccurate representation of the modeled eigenfunctions ($|\beta_q| \leq n_\lambda/2$) by the expansion and testing functions. The second term represents aliasing error due to the unmodeled modes of $\mathcal{L}$. Since the unmodeled modes are aliased by discretization to lower-order, modeled modes, the eigenvalues of unmodeled modes perturb the eigenvalues of the moment matrix and cause an additive spectral error term.

5.1.4 Spectral Error for Low-Order Basis Functions

We now specialize the treatment to the same piecewise polynomial basis functions used to discretize the EFIE for the cylinder in Section 3.1.3. In this case, the window function $T(-\beta)F(\beta)$ becomes $s^b(\beta)$, where

$$s(\beta) = \frac{\sin(\pi\beta/n_\lambda)}{\pi\beta/n_\lambda} \tag{5.18}$$

The basis order index b is defined in (3.26) to be $p + p' + 2$, where p and p' are the polynomial orders of the testing and expansion functions, respectively.

Using (5.18) in (5.17), we obtain the spectral error

$$E_{q,b} \simeq \frac{\sin^b(\pi\beta_q/n_\lambda)}{(\pi\beta_q/n_\lambda)^b} - 1$$
$$+ \frac{j\sqrt{1-\beta_q^2}}{n_\lambda} \frac{\sin^b(\pi\beta_q/n_\lambda)}{\pi^b} \sum_{s\neq 0} \frac{(-1)^{bs}}{(\beta_{q,s}/n_\lambda)^b\sqrt{(\beta_{q,s}/n_\lambda)^2 - 1/n_\lambda^2}} \tag{5.19}$$

For $1 \leq b \leq 3$, the projection error is dominant for small β_q/n_λ. Expanding the first term of (5.19) for large n_λ yields

$$E_{q,b}^{(1)} \simeq -\frac{b\pi^2\beta_q^2}{6n_\lambda^2} \tag{5.20}$$

for the projection error part of the total spectral error. This will determine the current solution error away from the singularities at the edge of the strip. For the scattering amplitude, the projection error cancels due to the variational property of the method of moments, and the aliasing error term with the summation over s determines the scattering amplitude error, but only for the contribution to the total scattering amplitude from the interior of the strip. The summation in the aliasing error term can be evaluated approximately for large n_λ, leading to results essentially identical to those obtained for the cylinder in (3.29) and (3.30) for pulse and triangle expansion functions.

5.1.5 The Magic "1/3" Discretization

Surprisingly, by making use of the spectral error analysis given above, it can be shown that accurate numerical results can be obtained using delta functions for both the expansion and testing basis sets. The order parameter for this discretization is $b = 0$. To avoid the singularity of the kernel, the testing and expansion functions cannot be located at the same points. We shift the testing functions symmetrically about each mesh element midpoint, so that $t(x) = h\delta(x + \alpha h/2)/2 + h\delta(x - \alpha h/2)/2$, where α is a parameter that adjusts the separation between the two delta functions. The expansion function is $f(x) = h\delta(x)$. The factor of h is included so that the delta functions have the same weight as the pulse and triangle functions.

Using the Fourier shift theorem, for the two delta testing functions $T(\beta) = \cos(\pi\beta_q/n_\lambda)$. For the expansion function, $F(\beta) = 1$. The spectral error is

$$E_{q,0} = \cos(\pi\beta_q/n_\lambda) - 1 + j\sqrt{1 - \beta_q^2} \sum_{s\neq0} \frac{\cos(\pi\beta_{q,s}/n_\lambda)}{\sqrt{\beta_{q,s}^2 - 1}} \tag{5.21}$$

Expanding both terms to second order for small β_q/n_λ and evaluating the sum over s leads to

$$E_{q,0} \simeq -\frac{2j}{n_\lambda} \ln\left[2\sin(\pi\alpha/2)\right] - \frac{\pi^2\alpha^2\beta_q^2}{2n_\lambda^2} - \frac{2j}{5n_\lambda^3} \tag{5.22}$$

For $\alpha = 0$, the leading term is infinite, which corresponds to divergence of the diagonal elements of the moment matrix when the testing and expansion points are the same. For $\alpha \neq 0$, the leading term has order n_λ^{-1}, which means that the numerical solution converges, but at a very slow rate.

While first-order convergence for delta expansion and testing functions is interesting, for a particular choice of the shift parameter, solution accuracy is even better. If $\alpha = 1/3$, the leading term of $E_{q,0}$ vanishes, and the error becomes considerably smaller. This special choice of basis yields a solution accuracy that is as good as combinations of pulse and triangle functions, with the significant advantage that numerical integration of moment matrix is not required—only evaluation of the kernel at the locations of the delta testing and expansion functions. It would be of great interest if a similar discretization could be developed for 3D scattering problems.

5.1.6 Quadrature Error

The above treatment assumed an ideal discretization and exact integration of moment matrix elements. The use of numerical quadrature to evaluate the integrals leads to an additional aliasing error component. Following the treatment in Section 3.3.1, we employ the M-point integration rule

$$\int_{-h/2}^{h/2} dx\, f(x) \simeq \sum_{n=1}^{M} f(\xi_n) w_n \tag{5.23}$$

The function $F(\beta)$ appearing in (5.13) is replaced by

$$F_M(\beta) = \frac{1}{h} \sum_{n=1}^{M} w_n f(\xi_n) e^{jk_0 \beta \xi_n} \tag{5.24}$$

$T(\beta)$ is modified similarly if quadrature is used for the testing integration.

The midpoint integration rule is given by $w_n = \delta = h/M$ and $\xi_n = (n - 1/2)\delta - h/2$. With this integration rule and the pulse expansion function, $F_M(\beta)$ becomes the periodic sinc function

$$F_M(\beta) = \frac{\sin(\pi\beta/n_\lambda)}{M \sin[\pi\beta/(Mn_\lambda)]} \tag{5.25}$$

This function has maxima at $\beta = sMn_\lambda$ for $s = 0, \pm1, \pm2, \ldots$. Inserting $F_M(\beta)$ in (5.17), expanding about the maxima, and evaluating the resulting sum over s leads to the spectral error

$$E_{q,1,M} \simeq -\frac{j\,2\ln 2}{Mn_\lambda} \tag{5.26}$$

Since $\lambda_q \simeq \eta/2$ for small q, this result is equivalent to (3.61). While this result was derived for pulse expansion functions, the order of the quadrature error is the same for other basis functions. Since this represents an additive aliasing error contribution that is larger in magnitude than the projection error, the superconvergence of the method of moments is destroyed by quadrature, as was found in Section 3.3.1 for the circular cylinder.

5.1.7 Current Solution Error

The current solution is singular at the edges of the strip, and for the TM polarization the singularity is not square integrable. While the main goal in this chapter is to understand the effect of numerical solution error near scatterer edges, we first need to understand the solution error for the interior of the strip, away from the edges.

We will refer to the interior region of the scatterer as $\tilde{C}$. This represents the scatterer with segments on the order of a wavelength in size removed at the edges. We will assume the scatterer is electrically large enough that the width of the interior region can be taken to be d. The current is bounded on $\tilde{C}$, so the L^2 norm and RMS value are finite and can be used to quantify the solution error.

We have seen that a Fourier function or plane wave type mode is an approximate eigenfunction of $\mathcal{L}$, and that a sampled plane wave mode is an approximate eigenvector of the moment matrix. Likewise, if the incident field is a plane wave, then the right-hand side of the linear system (2.27) is

$$
\begin{aligned}
V_n^i &= h^{-1} \int t_n(x) e^{jk_0 \cos \phi^{\text{inc}} x} \, dx \\
&= h^{-1} \int t(x - x_n) e^{jk_0 \beta x} \, dx \\
&= e^{jk_0 \beta x_n} T(\beta)
\end{aligned}
\tag{5.27}
$$

where $\beta = \cos \phi^{\text{inc}}$. To simplify the treatment, we will assume that the angle of incidence is such that $\cos \phi^{\text{inc}} = \beta_q$ for some q. In this case, the right-hand side is an approximate eigenvector of the same form as the transformation used in (5.9).

The eigenvalues of the moment matrix are estimated by $\hat{L}_{qq}$ in (5.15), from which it follows that the eigenvalues of $\mathbf{Z}^{-1}$ are approximately $\hat{L}_{qq}^{-1}$. Since (5.27) is an approximate eigenvector of the moment matrix, the surface current solution vector can be estimated as

$$
I_n \simeq \hat{L}_{qq}^{-1} T(\beta_q) e^{jk_0 \beta_q x_n}
\tag{5.28}
$$

which is similar to the physical optics approximation for the current solution, except that the amplitude of the physical optics current mode includes discretization error. This result implies that solution error for a flat scatterer surface is dominated by error in the physical optics mode.

The relative current solution error is

$$
\frac{\|\hat{J} - J\|_{\text{RMS}(\tilde{C})}}{\|J\|_{\text{RMS}(\tilde{C})}} \simeq \left| \frac{\hat{L}_{qq} - L_{qq} T(\beta_q)}{\hat{L}_{qq}} \right|
\tag{5.29}
$$

where $\text{RMS}(\tilde{C})$ denotes the norm defined in (2.51) for the interior part of the strip with small regions near the strip edges omitted from the error computation. By making use of the definition of $E_{q,b}$, and assuming that the spectral error is small, the relative RMS error becomes

$$
\text{Err}_{\text{RMS}(\tilde{C})} \simeq \left| E_{q,b} + 1 - T(\beta_q) \right|
\tag{5.30}
$$

Since $T(\beta_q) = T(-\beta_q)$ for symmetric testing functions, the term $1 - T(\beta_q)$ eliminates the smoothing error due to the testing functions and reduces the smoothing error term of (5.17) to approximately $F(\beta_q) - 1$. If the testing functions are delta functions located at the node points x_n, then $T(\beta_q) = 1$ and the current error is

$$
\text{Err}_{\text{RMS}(\tilde{C})} \simeq |F(\beta) - 1|
\tag{5.31}
$$

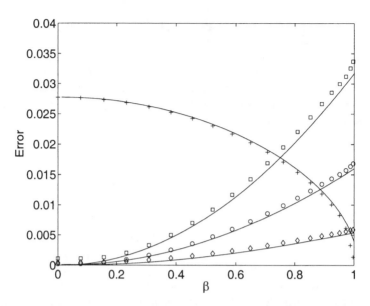

Figure 5.2: Relative RMS current error for a plane wave incident at an angle of $\phi^{\text{inc}} = \cos^{-1}\beta$ on a flat PEC strip. Integral equation: EFIE. Polarization: TM. Scatterer: flat strip, width $d = 10\,\lambda$. Mesh: flat-facet, $n_\lambda = 10$. Error is computed over 8λ interior region of the strip (edge singularities excluded). Diamonds: $b = 0$ ("1/3" discretization). Pluses: $b = 1$ (point testing, pulse expansion functions, single-point integration rule). Circles: $b = 1$ (point testing, pulse expansion functions, high-order integration rule). Squares: $b = 2$ (point testing, triangle expansion functions). Solid lines: theoretical estimate (5.31) and the single-point integration rule estimate from (3.64). (©John Wiley & Sons [2].)

where we have dropped the subscript q on β with the understanding that $\beta = \cos\phi^{\text{inc}}$, and we have assumed an ideal discretization, so the aliasing error is small and can be neglected.

For piecewise polynomial expansion functions of order p', using (5.18) this becomes

$$\text{Err}_{\text{RMS}(\bar{C})} \simeq \left| \left[\frac{\sin\left(\pi\beta/n_\lambda\right)}{\pi\beta/n_\lambda} \right]^{p'+1} - 1 \right| \tag{5.32}$$

For pulse expansion functions, $p' = 0$, and the relative RMS current solution error is

$$\text{Err}_{\text{RMS}(\bar{C})} \simeq \frac{\pi^2 \cos^2\phi^{\text{inc}}}{6n_\lambda^2} \tag{5.33}$$

which is identical to (2.68) with $k = k_0 \cos\phi^{\text{inc}}$. This result shows that the current solution error is second order in n_λ^{-1}, with the error caused by edge singularities excluded.

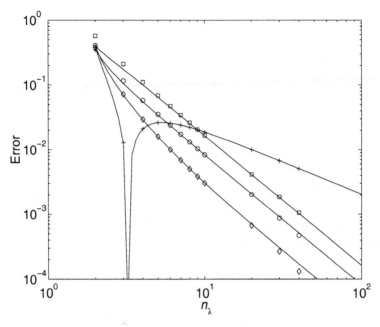

Figure 5.3: Relative RMS current error as a function of discretization density for a plane wave incident at an angle of $\phi^{inc} = \pi/4$. Integral equation: EFIE. Polarization: TM. Scatterer: flat strip, width $d = 10\,\lambda$. Error is computed over 8λ interior region of the strip (edge singularities excluded). Diamonds: $b = 0$ ("1/3" discretization). Pluses: $b = 1$ (point testing, pulse expansion functions, single-point integration rule). Circles: $b = 1$ (point testing, pulse expansion functions, high-order integration rule). Squares: $b = 2$ (point testing, triangle expansion functions). Solid lines: theoretical estimate (5.31) and the single-point integration rule estimate from (3.64). (©John Wiley & Sons [2].)

Figure 5.2 shows the relative RMS current error on the interior of a flat strip for a plane wave incident field as a function of the angle of incidence. The discretization density is fixed at $n_\lambda = 10$. A reference solution is obtained using a discretization density of $n_\lambda = 100$ with geometrical h-refinement near the edges of the strip. It can be seen that the single current mode approximation used in (5.28) leads to an accurate solution error estimate. For most of the discretizations, the error is smallest for broadside incidence, because the physical optics component of the current is constant along the scatterer surface and the projection error vanishes for a constant (DC) mode.

Figure 5.3 shows the relative RMS current error on the interior of the strip for a fixed incidence angle as a function of discretization density. The reference solution is obtained using a discretization density of $n_\lambda = 140$ with geometrical h-refinement at the edges of the strip. The error is asymptotically first order in n_λ^{-1} for the $b = 1$ case with a single-point integration rule, as predicted by (5.26). The error is second order for the other discretization schemes with high-order moment matrix integration rules.

5.1.8 Scattering Amplitude Error

The scattered field used in the computation of the scattering amplitude according to (2.45) is a plane wave traveling away from the scatterer at some angle ϕ^{sca}. By applying the derivation of (5.27) to (2.46), the discretized scattered plane wave has elements given by

$$V_n^s = e^{-jk_0\beta x_n} F(\beta) \tag{5.34}$$

where $\beta = \cos\phi^{sca}$. To simplify the treatment, we will consider the specular scattering amplitude, for which $\phi^{sca} = \pi - \phi^{inc}$. In this case, $\cos\phi^{sca} = -\cos\phi^{inc}$, and $V_n^s = V_n^i$.

Since V_n^s and V_n^i are approximate eigenvectors of the moment matrix, in the first-order scattering approximation, the numerical solution for the specular scattering amplitude due to the interior of the strip is

$$\hat{S} \simeq \frac{T(\beta_q)F(\beta_q)}{\hat{\lambda}_q} \tag{5.35}$$

From this approximation, we obtain the relative error

$$\frac{|S - \hat{S}|}{|S|} \simeq \left| E_q^{(2)} \right| \tag{5.36}$$

where $E_q^{(2)}$ is the second term (aliasing error) of (5.17) and S is the exact scattering amplitude. As with the circular cylinder, due to the variational property of the method of moments, the smoothing error does not contribute to the error in the scattering amplitude. We will defer numerical results for the scattering amplitude error until the edge current singularity contribution has been analyzed in Section 5.3 and can be combined with (5.36) to obtain a scattering amplitude error estimate for the full strip including edge effects.

5.2 Flat Strip Interior Error, TE-EFIE

As in the TM case, $\mathcal{N}$ can be decomposed into a normal approximation $\mathcal{H}$ and a non-normal perturbation $\mathcal{R} = \mathcal{N} - \mathcal{H}$. The normal approximation will be the diagonal part of a Fourier representation N_{qr} of the TE-EFIE operator $\mathcal{N}$. The diagonal elements N_{qq} provide approximations to the eigenvalues of $\mathcal{N}$ and will be used to analyze the error of method of moments solutions.

For the TE polarization, currents vanish at the edges of the strip. Accordingly, we employ sine and cosine functions rather than the complex exponentials in (5.1) to obtain a normal operator approximation. The Fourier representation of the TE-EFIE operator is

$$N_{qr} = d^{-1}c_q c_r \left\langle \cos\left(\beta_q k_0 x\right), \mathcal{N} \cos\left(\beta_r k_0 x\right) \right\rangle \qquad (5.37)$$

for even q and r. For odd q or r, the corresponding cosine function is replaced by a sine function. c_q is a normalization constant such that the transformation from the continuous operator to the Fourier representation is unitary, and we have redefined β_q to be $q/(2D)$. While the sine and cosine functions do not vanish at the strip edges with the same asymptotic behavior as the current solution, more accurate results are obtained than if the Fourier functions used in (5.37) were nonzero at the edges.

For the even modes, the diagonal elements are

$$N_{qq} = \frac{\eta}{2\pi^2 D} \int_{-\infty}^{\infty} d\beta \, \sin^2\left(\pi D\beta\right)\sqrt{1-\beta^2} \left[\frac{1}{(\beta-\beta_q)^2} - \frac{1}{(\beta-\beta_q)(\beta+\beta_q)}\right] \qquad (5.38)$$

For the odd modes, $\sin^2\left(\pi D\beta\right)$ is replaced by $\cos^2\left(\pi D\beta\right)$ in this expression. In [2], the expansion

$$N_{qq} \sim \frac{\eta\sqrt{1-\beta_q^2}}{2} + \frac{\eta}{2\pi^2}\left[j + \frac{\ln\left(-j\beta_q + \sqrt{1-\beta_q^2}\right)}{\beta_q\sqrt{1-\beta_q^2}}\right]D^{-1} + O(D^{-2}), \quad D \to \infty \qquad (5.39)$$

is derived for $\beta_q \neq 1$. This result is valid for both odd and even modes. For $\beta_q = 1$,

$$N_{qq} \sim \frac{\eta\sqrt{2}(1-j)}{2\pi}D^{-1/2} + \frac{j\eta}{\pi^2}D^{-1} + O(D^{-3/2}), \quad D \to \infty \qquad (5.40)$$

Since $\beta = 1$ corresponds to the surface wave mode with spatial frequency k_0, this is an approximation to the self-interaction of the surface wave mode.

From (5.39) and (5.40), it can be seen that up to a factor of $\eta/4$, the spectrum of the EFIE for the TE polarization is approximately the inverse of the spectrum of the TM-EFIE. This is related to the fact that the product of the two operators is a compact perturbation of the identity [4] and the Calderon identities associated with the integral operators of electromagnetics [5].

A numerical example of the spectrum of the TE-EFIE moment matrix for a flat strip is shown in Figure 5.4. Computed eigenvalues are compared to the spectral estimates (5.39) and (5.40). Starting with the zero spatial frequency or DC mode, the spectrum begins at $\eta/2$ on the real axis, which is the value of (5.39) for $\beta_q = 0$. As the modal spatial frequency increases, the eigenvalues move along the real axis with

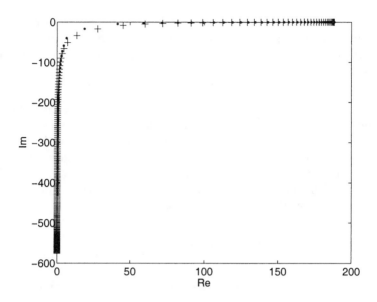

Figure 5.4: Spectrum of the moment matrix for the TE-EFIE for a flat strip, $D = 10$. Basis: point testing, pulse expansion functions. Mesh: flat-facet, $n_\lambda = 10$. Moment matrix integration: exact. Pluses: computed eigenvalues. Dots: eigenvalue estimates given by the normal operator approximation, (5.39) and (5.40).

increasing negative imaginary part. The eigenvalue of the surface wave mode given by (5.40) is closest to the origin. Modes with spatial frequency greater than k_0 have eigenvalues with very large magnitude along the negative imaginary axis. The increasing magnitude of the eigenvalues of highly oscillatory modes is a manifestation of the antismoothing or differentiating property of the TE-EFIE.

5.2.1 Discretized Operator Spectrum

Now that we have eigenvalue estimates for the continuous operator, we will develop estimates of the moment matrix eigenvalues, in order to determine the spectral error caused by operator discretization. The moment matrix elements for the TE-EFIE can be expressed as

$$Z_{mn} = \frac{\eta}{2n_\lambda} \int_\infty^\infty d\beta \sqrt{1 - \beta^2} e^{-j2\pi\beta(m-n)/n_\lambda} T(-\beta) F(\beta) \qquad (5.41)$$

where $T(\beta)$ and $F(\beta)$ are Fourier transforms of the testing and expansion functions as before.

To transform the moment matrix to a strongly diagonal matrix from which eigenvalue estimates can be obtained, we use a discrete version of the unitary transformation in (5.37). Applying this transformation to (5.41) leads to a matrix with elements

$$\hat{N}_{qr} = \frac{\eta}{2n_\lambda^2 D} \int_{-\infty}^{\infty} d\beta \sqrt{1 - \beta^2} B_q(\beta) B_r(\beta) T(-\beta) F(\beta) \tag{5.42}$$

The function $B_q(\beta)$ in this expression is

$$B_q(\beta) = c_q' \left[\frac{\sin[\pi D(\beta - \beta_q)]}{\sin[\pi(\beta - \beta_q)/n_\lambda]} - (-1)^q \frac{\sin[\pi D(\beta + \beta_q)]}{\sin[\pi(\beta + \beta_q)/n_\lambda]} \right] \tag{5.43}$$

where c_q' is a normalization constant. The diagonal elements of $\hat{N}_{qr}$ will provide eigenvalue estimates for the moment matrix.

5.2.2 Spectral Error

Using the estimates obtained above for the TE-EFIE operator eigenvalues and the moment matrix eigenvalues, the relative spectral error can be estimated as

$$E_q = \frac{\hat{\lambda}_q - \lambda_q}{\lambda_q} \simeq \frac{\hat{N}_{qq} - N_{qq}}{N_{qq}} \tag{5.44}$$

Since $B_q(\beta)$ is a combination of Dirichlet functions, the integrand of (5.42) is periodic in β. By expanding the integrand at each peak of $B_q(\beta)$, an approximation for the integral in (5.42) can be obtained. Following the derivation of (5.19) for the TM polarization, the relative spectral error for piecewise polynomial basis functions obtained from this procedure is

$$E_{q,b} \simeq T(-\beta_q) F(\beta_q) - 1 - \frac{j}{\sqrt{1 - \beta_q^2}} \sum_{s \neq 0} T(-\beta_{q,s}) F(\beta_{q,s}) \sqrt{\beta_{q,s}^2 - 1} \tag{5.45}$$

where $\beta_{q,s} = \beta_q + sn_\lambda$. This expression is similar to (5.17) for the TM case.

For point testing and pulse expansion functions ($b = 0$), the spectral error is nearly the same as that obtained in Section 3.4.2 for the circular cylinder. Because the TE-EFIE operator includes derivatives, one might expect that triangle expansion functions would be required to obtain meaningful results. As noted in Chapter 3, the series in (5.45) is not absolutely convergent, which is associated with instability of the discretization. If the testing points are not located exactly in the centers of the pulse expansion functions, the solution does not converge.

For point testing and triangle expansion functions or pulse testing and expansion functions ($b = 1$), the diagonal elements of the moment matrix diverge. In this case, the shifted symmetric testing approach of Section 5.1.5 can be used to obtain an accurate solution [2].

The lowest-order discretization for the TE polarization for which the series in (5.45) is absolutely convergent is pulse testing with triangle expansion functions ($b = 3$). For this choice of basis, the Fourier transforms of the testing and expansion functions decay as q^{-3}. Since the eigenvalues of the TE-EFIE operator increase as q, the terms in the series in (5.45) fall off as q^{-2}, which means that the series is absolutely convergent and the discretization is stable with respect to variations in the locations of the testing and expansion functions. The spectral error for that flat strip is similar to (3.76), with a second-order projection error term and a third-order aliasing error contribution.

5.2.3 Current Solution Error

Unlike the TM polarization, for the TE case the L^2 norm of the current solution exists over the entire strip, since the current is finite as long as there are no sources located on the strip itself. The current is singular, but the singularity is weaker than for the TM polarization and has the form $J_t(x) \sim x^{1/2}$, $x \to 0$ near the edges of the strip.

Although the current is finite for the TE polarization, the single-mode approximation of Section 5.1.7 for the TM polarization is not valid. Because the eigenvalue of the surface wave mode is small for the TE polarization, the amplitude of the corresponding component in the current solution is amplified by inversion of the TE-EFIE operator. As a result, the TE polarization has a strong surface wave mode. The amplitudes of modes with spatial frequencies near k_0 are larger relative to the dominant physical optics mode than is the case for the TM polarization, and these modes influence the current solution error.

Since the leading projection error component of the spectral error (5.45) is second order for low-order polynomial basis functions, the current error for all modes is second order, so the spectral error can still be used to obtain a reasonable current error estimate, in spite of the strong surface wave mode. As for the circular cylinder, the testing component of the projection error term of (5.45) cancels on a mode-by-mode basis, and the current solution error for the physical optics mode is identical to (5.31) for the TM polarization. For triangle expansion functions, the relative RMS current solution error is

$$\text{Err}_{\text{RMS}(\tilde{C})} \simeq \frac{\pi^2 \cos^2 \phi^{\text{inc}}}{3n_\lambda^2} \tag{5.46}$$

Figure 5.5 shows the relative RMS current solution error, with small regions at the edges of the strip omitted from the error computation to avoid the edge singularities of the current. A reference solution is obtained using a discretization density of $n_\lambda = 140$ with geometrical h-refinement at the strip edges.

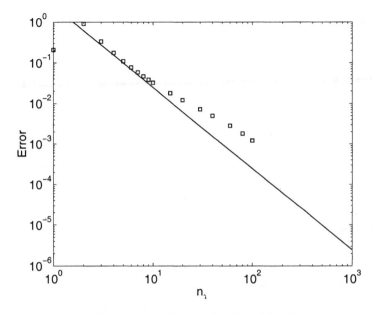

Figure 5.5: RMS current solution error for a flat strip of width $20\,\lambda$ for the TE-EFIE. Discretization: pulse testing and triangle expansion functions. Moment matrix element integrations: exact. Squares: numerical results. Solid line: theoretical error estimate. Error is computed over the interior of the strip with a one-wavelength region at each end of the strip excluded. (©2004 IEEE [6].)

5.2.4 Quadrature Error

For the point testing and pulse expansion ($b = 0$) and point testing with triangle expansion function ($b = 1$) discretizations, the contribution to the moment matrix elements of the hypersingular term in (2.11) can be evaluated analytically and numerical integration is not needed for that term. The remaining term that must be integrated numerically has the same kernel as for the TM polarization. For these lower-order discretizations, the error introduced by numerical quadrature is therefore the same as the error given by (5.26) in the TM case for $b \leq 1$.

For $b > 1$, the hypersingular term requires numerical integration. As was shown in Section 3.4.3, the spectral error can be reduced if the singularity of the integrand is weakened by integrating by parts. After integration by parts, the moment matrix elements are given by

$$Z_{mn} = \frac{\eta}{2n_\lambda} \int_{-\infty}^{\infty} \frac{d\beta}{\sqrt{1-\beta^2}} e^{-j2\pi k(m-n)/n_\lambda} \left[T(-\beta)F(\beta) - \frac{T'(-\beta)F'(\beta)}{k_0^2} \right] \quad (5.47)$$

where $T'(\beta)$ and $F'(\beta)$ are the Fourier transforms of the derivatives $t'(x)$ and $f'(x)$

of the testing and expansion functions, scaled by a factor of $1/h$. The Fourier representation of the moment matrix is then

$$\hat{N}_{qr} = \frac{\eta}{2n_\lambda^2 D} \int_{-\infty}^{\infty} \frac{d\beta}{\sqrt{1-\beta^2}} B_q(\beta) B_r(\beta) \left[T(-\beta) F(\beta) - \frac{T'(-\beta) F'(\beta)}{k_0^2} \right] \quad (5.48)$$

where the diagonal elements provide spectral estimates for the moment matrix.

As before, the effect of numerical integration on the moment matrix spectrum can be taken into account in (5.48) by evaluating the integrals in the Fourier transforms $T(\beta)$, $F(\beta)$, $T'(\beta)$, and $F'(\beta)$ of the testing and expansion functions and their derivatives using the quadrature rule. For the discretization with smoothness index $b = 3$ and an M-point integration rule with weights $w_n = \delta = h/M$ and abscissas $\xi_n = (n - 1/2)\delta - h/2$, we have

$$T_M(\beta) = \frac{\sin(\pi\beta/n_\lambda)}{M \sin[\pi\beta/(Mn_\lambda)]}$$

$$F_M(\beta) = \cos[\pi\beta/(Mn_\lambda)] \left(\frac{\sin(\pi\beta/n_\lambda)}{M \sin[\pi\beta/(Mn_\lambda)]} \right)^2$$

$$T'_M(k) = -2jh^{-1} \sin(\pi\beta/n_\lambda)$$

$$F'_M(k) = -\frac{2j \sin^2(\pi\beta/n_\lambda)}{hM \sin[\pi\beta/(Mn_\lambda)]}$$

The Fourier transforms of the derivatives of the basis functions are imaginary, because the derivatives are odd functions.

The spectral error introduced by the quadrature rule can be approximated by expanding the integrand of (5.48) around maxima at $\beta = sMn_\lambda$ for $s = \pm1, \pm2, \ldots$ and retaining the leading term for each s. This procedure leads to the spectral error contribution

$$E_{q,3,M} = \frac{j}{Mn_\lambda \sqrt{1-\beta_q^2}} \sum_{s \neq 0} \frac{(-1)^s \left[1 - \beta_q^2(-1)^{sM} \right]}{\sqrt{\left(s + \frac{\beta_q}{Mn_\lambda} \right)^2 - \left(\frac{1}{Mn_\lambda} \right)^2}} \quad (5.49)$$

For small β_q, the summation can be evaluated analytically, and the resulting error is identical to (5.26) for the TM polarization. From this, we can conclude that the solution error caused by numerical integration for pulse testing and triangle expansion functions is first order in n_λ^{-1}, and that integration by parts makes the behavior of the TE-EFIE with respect to numerical integration equivalent to that of the TM-EFIE.

5.3 Edge Error Analysis

In the preceding sections, we analyzed the solution and scattering amplitude error on the interior or smooth part of the scatterer, excluding the edge singularities of the geometry. To complete the error analysis for the flat strip, we now study the error at the edges of the strip caused by inaccurate representation of the current singularity. We will treat the case of the TM polarization, since the singularity is strongest and the edge effect has the greatest impact on the scattered fields. Since the current is not square integrable near the edges, we will measure the solution error in terms of the scattering amplitude.

By introducing the adjoint equation $\mathcal{L}^a J^a = E^s$, we can write the scattering amplitude in variational form as [7,8]

$$S = -\frac{k_0 \eta}{4} \frac{\langle E^s, J \rangle \langle J^a, E^i \rangle}{\langle J^a, \mathcal{L}J \rangle} \tag{5.50}$$

By substituting the approximate currents $\hat{J} = J + \Delta J$ and $\hat{J}^a = J^a + \Delta J^a$, the leading scattering amplitude error is found to be

$$\Delta S \simeq \frac{k_0 \eta}{4} \left(\langle \mathcal{L}\Delta J, \Delta J^a \rangle - \frac{\langle \Delta J, E^s \rangle \langle \Delta J^a, E^i \rangle}{\langle E^s, J \rangle} \right) \tag{5.51}$$

which at least formally is second order in the solution errors ΔJ and ΔJ^a. This expression provides a more convenient way to estimate scattering amplitude error than (3.57), since the second term in (5.51) does not require that the operator $\mathcal{L}$ be applied to the solution error ΔJ. This approach relies on the result that the scattering amplitude computed by (5.50) using a moment method solution for the current is identical to that computed with the definition (2.44) [8].

The first-order error factor $\langle \Delta J, E^s \rangle$ in (5.51) can be estimated by assuming the quasioptimality of $\hat{J}$ in the sense of Section 1.3.4. For a piecewise constant expansion,

$$\hat{J}(x) = \sum_n J(x_n) f(x - x_n) \tag{5.52}$$

where $f(x)$ is the pulse function given by (2.30b), $x_n = (n - 1/2)h$, and here x is the distance from the edge of the strip. If the amplitude of the incident field is unity, then by the results of Section 5.1, the amplitude of the current away from the edge is approximately Λ_q^{-1}, where $q = D \cos \phi^{\text{inc}}$. The current at the edge can then be approximated by

$$J(x) \simeq \Lambda_q^{-1} (x/\lambda)^{-1/2} \tag{5.53}$$

By making use of this approximation, the first-order edge error can be estimated as

$$\langle \Delta J, E^s \rangle \simeq \Lambda_q^{-1} h \sum_{n=1}^{n_\lambda} (x_n/\lambda)^{-1/2} - \Lambda_q^{-1} \int_0^\lambda dx \, (x/\lambda)^{-1/2} \tag{5.54}$$

It can be shown that

$$N^{-1/2} \sum_{n=1}^{N} (n - 1/2)^{-1/2} \sim 2 - cN^{-1/2}, \quad N \to \infty$$

where $c = \zeta(2)(1 - \sqrt{2}) \simeq 0.6$. The integral on the right of (5.54) evaluates to 2λ. The first-order error is finally

$$\langle \Delta J, E^s \rangle \simeq 0.6 \, \Lambda_q^{-1} \lambda n_\lambda^{-1/2} \tag{5.55}$$

which is of order $h^{1/2}$ with respect to the mesh element length.

We now assume that the first term on the right of (5.51) is on the order of the second term or smaller, and that the solution error for the adjoint equation is similar to that of the original integral equation. This leads to the edge error contribution

$$\Delta S = |S - \hat{S}| \simeq (k_0 \eta / 4) 0.2 \Lambda_q^{-1} \lambda n_\lambda^{-1} \tag{5.56}$$

This result indicates that the current error at the strip edges produces a first-order scattering amplitude error contribution.

To combine the edge error (5.56) with the interior error (5.36), we must normalize (5.56) by the magnitude of the scattering amplitude. Since this is not available in closed form, we will use the specular scattering approximation

$$S \simeq (k_0 \eta / 4) \Lambda_q^{-1} (d + 4\lambda)$$

Adding the two error contributions and assuming that d is large leads to the relative specular scattering amplitude error estimate

$$\frac{|S - \hat{S}|}{|S|} \simeq \left| E_q^{(1)} + 0.4 \, D^{-1} n_\lambda^{-1} \right| \tag{5.57}$$

For pulse expansion functions ($b = 1$), the error estimate becomes

$$\frac{|S - \hat{S}|}{|S|} \simeq \left| 1.8 \, j \eta \beta_q^2 \Lambda_q^{-1} n_\lambda^{-3} + 0.4 \, D^{-1} n_\lambda^{-1} \right| \tag{5.58}$$

where $\Lambda_q \simeq (\eta/2)/\sqrt{1 - \beta_q^2}$ and $\beta_q = \cos \phi^{\text{inc}}$. The first term is the third-order error contribution from spectral aliasing error associated with the interior of the strip, and the second term is a first-order error due to the edge singularities of the current solution.

As the mesh is refined and n_λ becomes large, the first-order edge contribution dominates the scattering amplitude error (5.57). For an electrically large scatterer, the

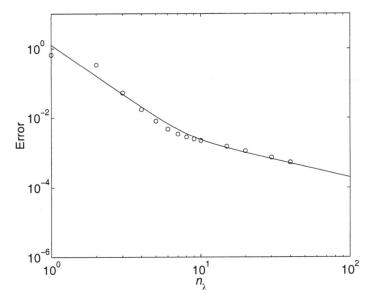

Figure 5.6: Relative specular scattering amplitude error for a flat PEC strip of width 20λ for the TM-EFIE. Incidence angle: $\phi^{inc} = \pi/4$. Scattering angle: $\phi^{sca} = 3\pi/4$. Discretization: point testing, pulse expansion. Circles: numerical results. Line: theoretical error estimate (5.57). (©2004 IEEE [6].)

scattering contribution from the interior part of the strip in the specular direction is stronger than the edge diffraction. This accounts for the $1/D$ dependence in the second term of (5.57). For nonspecular scattering directions, the scattering from the interior of the strip is smaller, and error due to edge effects is more important.

Figure 5.6 compares the estimate (5.58) to the computed solution error for a flat strip of length 20λ. A reference solution is obtain using a mesh density of $n_\lambda = 200$. Figure 5.7 shows the theoretical error estimate (5.57) for several different discretizations, along with numerical results. It can be seen that the asymptotic error is first order in n_λ^{-1}, regardless of the choice of basis functions or moment matrix integration rule. From these results, it can be seen that unless moment matrix elements are not integrated accurately and quadrature error is significant, edge current singularities are the dominant error contribution. By the quasioptimality principle, edge error is more sensitive to the choice of expansion functions than to the integral operator, so the first-order error estimate extends to the TM-MFIE and TM-CFIE integral formulations.

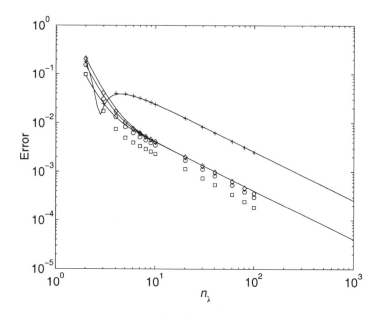

Figure 5.7: Relative specular scattering amplitude error for a flat PEC strip of width $10\,\lambda$ for the TM-EFIE. Incidence angle: $\phi^{\text{inc}} = \pi/3$. Scattering angle: $\phi^{\text{sca}} = 2\pi/3$. Diamonds: "1/3" discretization. Pluses: $b = 1$ (point testing, pulse expansion, single-point integration rule). Circles: $b = 1$ (point testing, pulse expansion, high-order integration rule). Squares: $b = 2$ (point testing, triangle expansion). Lines: theoretical error estimates.

5.4 WEDGES

The flat strip is an extreme case of a singular geometry for which the error due to inaccurate modeling of the singular edge currents leads to a first-order and typically dominant solution error contribution. We now wish to determine the solution error for a less extreme geometrical singularity, the PEC wedge.

In order to avoid other geometrical singularities, the natural scatterer geometry for this analysis is a conecircle, for which a wedge is joined in a smooth way to a circular cylinder as shown in Figure 5.8. To obtain an error estimate for this scatterer, we will combine the circular cylinder error estimate

$$|S - \hat{S}| \simeq 1.8 n_\lambda^{-3} \qquad (5.59)$$

obtained from the results of Chapter 3 with an estimate for the error due to the singular behavior of the surface current at the wedge tip.

The treatment of Section 5.3 can be readily applied to the case of a wedge by chang-

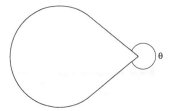

Figure 5.8: Geometry of the conecircle scatterer.

ing the singularity of the current solution in (5.53) to

$$J(x) \sim x^{-\gamma} \tag{5.60}$$

where $\gamma = 1 - \pi/\theta$ and θ is the exterior angle of the wedge. The resulting scattering amplitude error estimate is

$$|\hat{S} - S| \simeq 1.8\, n_\lambda^{-3} + C(\gamma)(k_0 a n_\lambda)^{-2(1-\gamma)} \tag{5.61}$$

where $C(\gamma) = (1 - \gamma)(2^\gamma - 1)\zeta(\gamma)$ and ζ is the Riemann zeta function.

The exponent of the wedge singularity error term in (5.61) ranges from −1 at $\theta = 360°$, where the conecircle collapses to a flat strip, to −2 at $\theta = 180°$, for which the scatterer is a smooth circular cylinder. In the latter case, $C(0) = 0$, so that the error estimate reduces to (5.59) as expected. Of note is the jump of the error from second order to third order as the scatterer becomes smooth. This is related to the second-order error due to flat mesh facets observed in Section 3.3.3.

The theoretical error estimate (5.61) is compared to numerical results for the backscattering amplitude in Figure 5.9. Reference solutions are obtained using a mesh density of $n_\lambda = 1,000$. It can be seen that there is good agreement between theory and numerical results.

To illustrate that the same error trends hold for bistatic scattering, Figure 5.10 shows numerical results for the maximum relative scattering width error over a range of bistatic scattering angles. This error metric also compares well to the theoretical estimate (5.61). Other error measures also behave similarly, including the scattering width error at backscattering, the RMS scattering width error over a range of scattered angles, and the RMS scattering width error in decibels.

5.5 Summary

From the results developed in this chapter, it can be seen that geometrical singularities lead to a significant solution error contribution. The total error consists of contributions from the smooth interior of the scatterer and the current singularities at edges

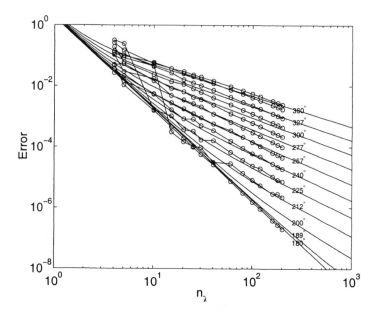

Figure 5.9: Backscattering amplitude error for the conecircle geometry for several exterior wedge angles ($\theta = 360°$ is a flat strip and $\theta = 180°$ is a circle). The distance from the circle center to the wedge tip is $\lambda/2$, and the angle of incidence is $\phi^{inc} = 60°$ relative to the scatterer horizontal axis. Integral equation: EFIE. Polarization: TM. Mesh: curved facets, varying mesh density n_λ. Discretization: point testing and pulse expansion functions. Circles: numerical results. Lines: theoretical error estimate. (©2004 IEEE [6].)

or corners. For far field quantities, the error due to edge effects for a flat strip with the TM polarization is first order in the mesh element width. For conducting wedges, the asymptotic error falloff rate decreases in proportion to the exterior angle of the wedge, so that the sharper the wedge, the larger the solution error at a given mesh density. At a mesh density of ten points per wavelength, relative error can range from worse than 10% for a sharp wedge to much better than 1% for a wedge with larger interior angle.

For electrically large scatterers, the contribution to far fields from smooth, flat regions of the scatterer may dominate diffracted fields from isolated geometrical singularities, so that error with respect to mesh density is initially smaller and converges more rapidly. As the mesh is refined, eventually the error contribution from the singularities will become dominant and the solution convergence rate reduces to a slower asymptotic value determined by the sharpness of the geometrical singularity.

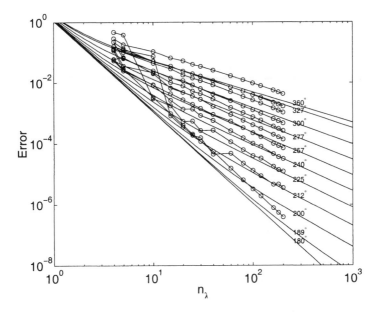

Figure 5.10: Maximum relative bistatic scattering width error over 31 equally spaced scattering angles for the same scatterer and method of moments implementation as in Figure 5.9. (©2004 IEEE [6].)

REFERENCES

[1] A. G. Ramm, "Eigenfunction expansion of a discrete spectrum in diffraction problems," *Radiotek. i Elektron.*, vol. 18, pp. 364–369, 1973.

[2] K. F. Warnick and W. C. Chew, "On the spectrum of the electric field integral equation and the convergence of the moment method," *Int. J. Numer. Meth. Engr.*, vol. 51, pp. 31–56, May 2001.

[3] S. C. Eisenstat and I. C. F. Ipsen, "Three absolute perturbation bounds for matrix eigenvalues imply relative bounds," *SIAM J. Matrix Anal. Appl.*, vol. 20, no. 1, pp. 149–158, 1998.

[4] S. Amini and S. M. Kirkup, "Solution of Helmholtz equation in the exterior domain by elementary boundary integral methods," *J. Comp. Phys.*, vol. 118, pp. 208–221, 1995.

[5] G. C. Hsiao and R. E. Kleinman, "Mathematical foundations for error estimation in numerical solutions of integral equation in electromagnetics," *IEEE Trans. Ant. Propag.*, vol. 45, pp. 316–328, Mar. 1997.

[6] K. F. Warnick and W. C. Chew, "Error analysis of the moment method," *IEEE Ant. Propag. Mag.*, vol. 46, pp. 38–53, Dec. 2004.

[7] R. F. Harrington, *Field Computation by Moment Method*. Malabar, FL: Krieger, 1982.

[8] D. S. Jones, "A critique of the variational method in scattering problems," *IRE Trans. Ant. Propag.*, vol. AP-4, no. 3, pp. 297–301, 1956.

Chapter 6

Resonant Structures

Resonance is a wave phenomenon associated with stored electromagnetic energy. Resonance can be characterized generally in terms of radiated power relative to the stored energy per period at a given frequency. A system with a high-quality factor (high-Q) resonance stores a large amount of energy while radiating only a small portion to the environment. For time harmonic problems, the system is assumed to be at a steady state, so the field solution represents a balance between supplied and radiated power. A three-dimensional structure has an infinite number of resonances, and so can be thought of as a combination of an infinite number of LCR circuits, each with its own resonant frequency and quality factor.

For exterior scattering from a PEC body, the scatterer is impenetrable and power is not coupled into internal cavities, so the interior of a hollow, closed scatterer has no physical impact on the external scattering behavior of the object. As far as the impact of resonance on a numerical method, however, internal resonances can affect solution accuracy significantly. For a conducting scatterer, an internal resonance is associated with a fictitious closed cavity resonator having the same shape as the scatterer. An internal resonance is nonphysical in the sense that even if the scatterer is hollow, no energy can enter the interior region from the outside. If the scatterer is a solid conductor, it does not have an internal cavity at all. Numerically, an internal resonance manifests itself as a zero eigenvalue of the EFIE operator, which leads to a small eigenvalue for the moment matrix and causes increased numerical error. The MFIE operator also can have zero eigenvalues, corresponding to internal resonances with a PMC type boundary condition.

Physical resonance is associated with concave regions of a scatterer such as a cavity, duct, or inlet. For a hollow scatterer with a small hole, at a resonance frequency of the structure the stored energy is large and the radiated power is small, so the quality factor of the resonance is high. If the opening of a cavity is large, the quality factor of the resonance is small. Even a scatterer with no concave region can have modes that might be viewed as weakly resonant or quasi-resonant.

The goal in this chapter is to analyze the impact of resonance on solution accuracy for the method of moments. We will consider both internal resonance and cavity resonance. In both cases, we will develop eigenvalue estimates for the EFIE operator for a resonant mode at or near the resonant frequency. Including discretization error using the approach of previous chapters will lead to spectral error expressions and current and scattering amplitude solution error estimates.

Normally, we think of resonant modes as electric and magnetic fields near or inside a resonant structure. Since the focus here is on surface integral equations, which deal with currents rather than fields, in this chapter "resonant mode" will refer to the currents induced on the scatterer by the modal fields, rather than the fields themselves.

6.1 RESONANCE AND THE EFIE OPERATOR SPECTRUM

We have already seen that the spectrum of the integral operators of electromagnetics determines in large part the solution accuracy of the method of moments. This close connection between the operator spectrum and numerical behavior continues to hold in the case of a resonant structure.

Using Poynting's theorem, we can relate the energy storage properties of a given eigenmode to the operator eigenvalue. For time-harmonic fields in free space, Poynting's theorem is [1]

$$-\int_V \mathbf{E} \cdot \mathbf{J}^* \, dv = \oint_{\partial V} (\mathbf{E} \times \mathbf{H}^*) \cdot d\mathbf{s} + j\omega \int_V \left(\mu|\mathbf{H}|^2 - \epsilon|\mathbf{E}|^2 \right) dv \tag{6.1}$$

where V is a region containing the scatterer with boundary surface ∂V. Considering a PEC scattering problem, we will apply this theorem to the equivalent surface current $\mathbf{J}_s$ on the scatterer surface S. When impressed in free space, this current radiates the fields $\mathbf{E}^{sca}$ and $\mathbf{H}^{sca}$. Using (2.4), Poynting's theorem becomes

$$\int_S \mathcal{T}\mathbf{J}_s \cdot \mathbf{J}_s^* \, dv = \oint_{\partial V} \mathbf{E}^{sca} \times \mathbf{H}^{sca*} \cdot d\mathbf{s} + j\omega \int_V \left(\mu|\mathbf{H}^{sca}|^2 - \epsilon|\mathbf{E}^{sca}|^2 \right) dv \tag{6.2}$$

The left-hand side of (6.1) changed sign because $\mathcal{T}\mathbf{J}_s$ is equal to the incident field on the scatterer surface, which by the PEC boundary condition is the negative of the scattered field radiated by the impressed current $\mathbf{J}_s$.

If $\mathbf{J}_s$ is an eigenfunction of $\mathcal{T}$, such that $\mathcal{T}\mathbf{J}_s = \lambda \mathbf{J}_s$, then we have

$$\lambda \int_S |\mathbf{J}_s|^2 \, dv = P_{rad} + j\omega(U_H - U_E) \tag{6.3}$$

where P_{rad} denotes the surface integral term of (6.2) and $U_H - U_E$ is the volume integral term. If the eigenfunction $\mathbf{J}_s$ is normalized to have unit L^2 norm, then the eigenvalue is

$$\lambda = P_{\text{rad}} + j\omega(U_H - U_E) \tag{6.4}$$

This result shows that the energy storage properties of a scatterer are closely related to the eigenvalues of the EFIE operator. The real part of the eigenvalue is the total power radiated by the current mode to the far field, and the imaginary part is determined by the difference between the magnetic and electric energies stored in the scattered field.

With a simple LCR circuit, there is a balance between the stored electric and magnetic energies at the resonance frequency. We will see shortly that the same holds for a resonant mode of a conducting structure. At the resonance frequency of the mode, the imaginary part of the eigenvalue (6.4) vanishes. If the frequency is slightly different from the resonant frequency of the mode, the stored electric and magnetic energies are not exactly equal, and the eigenvalue has a small positive or negative imaginary part. The radiated power is frequency-dependent, but the dependence is weaker than that of the imaginary part, so that as a function of frequency, the eigenvalue associated with a resonant or near resonant mode moves in a roughly vertical trajectory across the positive real axis in the complex plane. For a high Q resonance, little real power is radiated by the mode and the real part P_{rad} of the eigenvalue is small, so the magnitude of the eigenvalue $\lambda = P_{\text{rad}}$ is small at the resonance frequency and the eigenvalue follows a path that is very close to the imaginary axis.

For a closed scatterer, an internally resonant mode has an eigenvalue with zero real part ($P_{\text{rad}} = 0$) at the resonance frequency. As a function of frequency, the eigenvalue moves along the imaginary axis and passes through the origin at resonance, so that the eigenvalue is identically zero at resonance.

6.1.1 Quasi-Resonant Modes

Some scatterers have modes with eigenvalues that are small in magnitude and so can be considered as quasi-resonant, but the modes do not have all the characteristic behaviors associated with a resonant mode. For a flat strip with TE polarized fields, the eigenvalue of the TE-EFIE associated with the surface wave mode is small, and approaches zero as the electrical size of the scatterer becomes large. Because the eigenvalue has a small magnitude, the amplitude of the surface wave mode in the current solution is larger than for the TM polarization. As a function of frequency, however, the eigenvalue associated with the surface wave mode does not cross the real axis in the complex plane at a particular frequency.

For the TM-EFIE, highly oscillatory nonradiating modes that radiate only evanescent fields have eigenvalues with a small imaginary part and an even smaller real part. These modes are amplified in the current solution and cause edge current singularities. As is the case with the TE surface wave mode, these eigenvalues do not cross the real axis as the frequency is changed.

6.1.2 Resonance and the Method of Moments

The small operator eigenvalue associated with a resonant mode can significantly reduce the accuracy of the method of moments at or near the resonance frequency. We will use the spectral error defined in Chapter 3 as an analysis tool to quantify this effect. The spectral error in the moment matrix eigenvalue given by (3.15) consists of multiplicative and additive contributions. In a relative sense, a small eigenvalue is perturbed more by an additive shift than by a multiplicative scale factor. Consequently, the relative spectral error (3.18) is large for the small eigenvalue associated with a resonant mode. The task now is to analyze the impact of this large relative spectral error for internal and real resonances on current and scattering amplitude solution error.

6.2 INTERNAL RESONANCE

In order to analyze the internal resonances of a closed conducting scatterer, it is convenient to return to the canonical case of the circular PEC cylinder considered in Chapter 3. At an internal resonance frequency, one of the eigenvalues of the integral operator vanishes. For the TM-EFIE, in consideration of (3.2) the internal resonance condition is

$$J_q(k_0 a) = 0 \tag{6.5}$$

for some integer q. At a resonance frequency, $k_0 a = X$, where X is a zero of $J_q(x)$. If this condition holds, then the qth eigenmode on the cylinder is internally resonant.

This can be recognized as the governing relationship for the TM modes of a hollow circular cylindrical PEC waveguide. From a partial differential equation point of view, (6.5) is the condition for zero eigenvalues of the two-dimensional Laplace problem for a disk of radius a. When the interior Laplace problem with the Dirichlet boundary condition on the scatterer surface S has a nontrivial solution, the exterior TM-EFIE operator has a zero eigenvalue.

For the continuous electric field integral equation, it can be seen from (3.32) that the amplitude of the qth Fourier mode in the incident field on the right-hand side of the moment method linear system (2.8) approaches zero as $k_0 a \to X$. The amplitude of the qth mode in the right-hand side and the qth operator eigenvalue both vanish at the internal resonance frequency. In (3.34), it is apparent that the amplitude of the qth mode in the current solution is the amplitude of the mode in the right-hand side divided by the eigenvalue. This ratio has a finite limit at $k_0 a = X$. Therefore, in regard to the continuous TM-EFIE, the internal resonance is only mathematical and does not correspond to unusual physical behavior of the current mode.

When the continuous integral equation is discretized, the eigenvalue is shifted by the spectral error Δ_q defined in (3.16). Expanding the exact eigenvalue (3.2) about

$k_0 a = X$ leads to

$$\hat{\lambda}_q = -j\frac{\eta\pi X}{2}J'_q(X)Y_q(X)(k_0 a - X) + \Delta_q \qquad (6.6)$$

for the moment matrix eigenvalue. Although Δ_q consists of multiplicative and additive parts, as noted above the additive part of the spectral error is most important in determining the numerical behavior of the moment method at resonance frequencies.

Since the additive part of the spectral error is associated with aliasing of higher-order eigenvalues, which are nearly pure imaginary, the spectral error Δ_q is imaginary. In the complex plane, as a function of frequency the exact eigenvalue λ_q moves vertically along the imaginary axis. The moment matrix eigenvalue $\hat{\lambda}_q$ follows the same trajectory, but with a small vertical offset due to the imaginary spectral error. The approximate eigenvalue crosses the origin at a slightly different frequency than the exact eigenvalue. As a result, discretization error shifts the internal resonance. Although the real part is very small, the spectral error Δ_q is not purely imaginary, so the eigenvalue is shifted slightly away from the imaginary axis and does not become identically zero at the shifted resonance frequency.

The shift in resonance frequency in itself may not be a significant problem in most applications of the moment method. At the exact resonance frequency ($k_0 a = X$), however, the spectral error causes a significant error in the current solution. At resonance, the approximate eigenvalue (6.6) is nonzero due to the spectral error. The amplitude of the resonant mode in the incident field on the right-hand side of the EFIE is not affected by the additive aliasing error, and only includes a factor of T_{-q} due to the error associated with projection of the mode onto the testing functions. As a result, discretization shifts the operator resonance frequency but not the right-hand side resonance frequency [2]. The amplitude of the resonant mode in the numerical solution is the zero mode amplitude in the right-hand side divided by the spectral error Δ_q. Rather than taking on a finite limit as in the continuous case, numerically, the mode amplitude in the current solution is zero. As a result, the approximate current solution is missing the resonant mode. This leads to a very large current solution error at the exact resonant frequency.

Whether or not the amplitudes of other modes in the current solution are accurate even at the resonance frequency is a different question. This depends on how the linear system associated with the moment method is solved. At the exact resonance frequency, the moment matrix is nearly singular, due to the resonant eigenvalue (6.6), which is equal to the small spectral error Δ_q. Depending on the linear system solution method, the near singularity of the moment matrix can corrupt the amplitudes of other modes in the numerical current solution. As an example, the conjugate gradient (CG) iterative solver can be applied to the normal form

$$\mathbf{Z}^H\mathbf{Z}\mathbf{i} = \mathbf{Z}^H\mathbf{v}^i \qquad (6.7)$$

of the linear system. The amplitude of the resonant mode in the new right-hand side

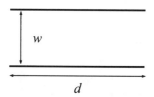

Figure 6.1: Parallel strip resonator.

is small, since the corresponding eigenvalue is small. Because this eigenvector is not present in the right-hand side, the behavior of the CG algorithm is governed by an effective matrix operator that no longer has the vanishing eigenvalue, and the iteration converges at a rate similar to nonresonant frequencies.

At frequencies near the exact resonance, the small eigenvalue associated with the resonant mode is nonzero, but is still strongly perturbed in a relative sense by spectral error, so the amplitude of the resonant mode in the numerical current solution is incorrect. Therefore, as a function of frequency there is a peak in current solution error near each resonant frequency. This is apparent (for the MFIE) in Figure 4.1. A similar behavior is encountered for the EFIE.

For scattered fields, the resonant modes associated with the EFIE do not radiate to the far field, whereas the resonant modes of the MFIE do radiate [2]. Therefore, for the EFIE scattered fields can be more accurate than for the MFIE. At the shifted resonance corresponding to the smallest value of the moment matrix eigenvalue (6.6), the linear system cannot be solved accurately and scattered fields are inaccurate. The scattering amplitude error is therefore large near the shifted resonance frequency, but the peaks in the error as a function of frequency for the scattering amplitude with the EFIE can be narrower than the peaks in the current solution error [2,3]. For the MFIE, the current and scattering amplitude errors are both large over a broad band, since the resonant mode radiates to the far field and contributes to the scattering amplitude near the resonance frequency.

6.3 CAVITIES

As noted in Chapter 2, numerical difficulties caused by internal resonance can be eliminated using the CFIE formulation, since the CFIE does not have small eigenvalues near the internal resonance frequencies of the EFIE or MFIE. A cavity resonance is physically coupled to external fields and the effects of resonance on the operator spectrum are manifested by the both the EFIE and MFIE (and the CFIE as well).

In order to analyze the numerical behavior of the method of moments for a cavity

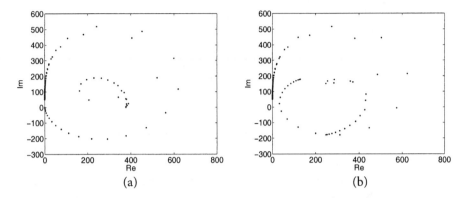

Figure 6.2: Spectrum of the TM-EFIE operator for the parallel strip resonator. (a) $D = 10$, $W = 1$. (b) $D = 10$, $W = 1.2$.

resonance, we will consider the canonical resonant scatterer consisting of two parallel PEC strips of width d, separated by a distance w. This is the parallel strip resonator, shown in Figure 6.1. We will also consider the half-open parallel strip resonator, for which one of the ends is closed by a side wall, so that the cavity is open only on one end.

Figure 6.2 shows the TM-EFIE spectrum for the parallel strip resonator for two different values of the electrical separation distance $W = w/\lambda$ between the two strips. In both cases, the depth in wavelengths is $D = d/\lambda = 10$. As with the circular cylinder shown in Figure 3.1 and the single PEC strip in Figure 5.1, the eigenvalues of nonradiating, rapidly oscillating modes approach the origin along the positive imaginary axis. These eigenvalues are insensitive to the global geometry of the scatterer. The eigenvalues of propagating modes depend strongly on the scatterer shape, and for the parallel strip resonator, eigenvalues of resonant modes are spaced along an arc that also approaches the origin. As we will see shortly, for a plate electrical separation of $W = 1$, the lowest-order mode of the cavity is close to resonance. Ignoring edge diffraction, the surface current associated with this mode is a half sinusoid on each strip that vanishes at the strip ends and peaks at the midpoints. In Figure 6.2(a), the eigenvalue corresponding to this mode is very close to the origin. The imaginary part is small due to the balance of electric and magnetic stored energies at resonance, and the real part is small due to the high quality factor of the resonant mode. For a different plate separation ($W = 1.2$), higher-order modes with more oscillations across the strips are close to resonance, but these modes have a lower quality factor and hence the eigenvalues have a larger real part and are farther from the origin.

Analytically, the advantage of the parallel strip resonator as a canonical scatterer is that the treatment of Chapter 5 for the PEC strip can be applied to obtain eigenvalue estimates. We will develop a normal operator approximation $\mathcal{L}$ with elements L_{qr} that represent the interactions between Fourier modes on the two strips. The matrix representation of the operator will have a block structure of the form

$$
\mathbf{L} = \begin{bmatrix} \mathbf{L}^s & \mathbf{L}^c \\ \mathbf{L}^{cH} & \mathbf{L}^s \end{bmatrix} \tag{6.8}
$$

The upper left block represents interactions between Fourier modes on the lower strip. The lower right represents interactions between modes on the upper strip. By symmetry, the two diagonal blocks are identical. The off-diagonal blocks represent fields received by Fourier modes on one strip due to radiation from modes on the other strip, or the cross-coupling between the two strips.

We will first consider the diagonal blocks of (6.8). The surface currents associated with the resonant modes of a closed rectangular cavity vanish at the endpoints of each cavity wall, so that we need only consider the Fourier modes that vanish at the ends of the strip. For the TM polarization, the interaction between two sinusoidal modes on one of the strips is

$$
L_{qr}^s = c_q c_r \frac{k_0 \eta}{4} \int_{-d/2}^{d/2} \int_{-d/2}^{d/2} dx\, dx'\, H_0^{(2)}(k_0|x - x'|) \sin(\beta_q k_0 x) \sin(\beta_r k_0 x') \tag{6.9}
$$

where $\beta_q = q/(2D)$ is the normalized spatial frequency of the mode of order q and c_q is a mode normalization constant. $D = d/\lambda$ is the electrical length of the cavity. This expression is given for q and r odd. If q or r is even, the corresponding sine function is replaced by the cosine function.

By making use of the 1D Fourier representation (5.2) of the operator kernel, (6.9) becomes

$$
L_{qr}^s = \frac{\eta}{2\pi^2 D} \int_{-\infty}^{\infty} \frac{d\beta}{\sqrt{1 - \beta^2}} B_q(-\beta) B_r(\beta) \tag{6.10}
$$

This is nearly identical to (5.3), except that because the resonant modes vanish at the strip ends, the current boundary condition at $x = \pm d/2$ is similar to that of the TE polarization for a single strip. B_q is therefore given by (5.43) rather than the periodic sinc function in (5.3).

When two identical physical systems are coupled, generally the spectrum of the governing operator for one of the systems is split by the coupling into pairs of sum and difference eigenvalues. This is the case for the parallel strip resonator. The eigenvalues of the normal operator approximation (6.8) consist of sums and differences of the self-coupling and cross-coupling. Since each block of (6.8) is strongly diagonal, the eigenvalues of the EFIE operator $\mathcal{L}$ for the cavity can be approximated by sums and differences of the diagonal elements of $\mathbf{L}^s$ and $\mathbf{L}^c$. The cross-coupling between modes on the two parallel strips can be found by modifying (5.3) to

$$L_{qr}^c = \frac{\eta}{2\pi^2 D} \int_{-\infty}^{\infty} \frac{d\beta}{\sqrt{1-\beta^2}} B_q(-\beta) B_r(\beta) e^{-j2\pi W \sqrt{1-\beta^2}} \tag{6.11}$$

The exponential accounts for the effects of field propagation from one strip to the other.

It is typical to analyze resonant phenomena as a function of frequency. Since we have represented the scatterer in terms of the dimensionless electrical lengths $D = d/\lambda$ and $W = w/\lambda$, changing the frequency amounts to changing D and W while keeping the aspect ratio D/W fixed. We will first consider the case of D and W such that one of the modes of the structure is exactly at resonance.

6.3.1 Resonant Case

At resonance, the difference eigenvalue for one of the modes of the parallel strip resonator is at a minimum, meaning that the diagonal elements of $\mathbf{L}^s$ and $\mathbf{L}^c$ associated with the mode are nearly equal. We can use this in conjunction with the normal operator approximation (6.8) to find a resonance condition for the parallel strip resonator.

By making use of (5.3) and (6.11), the difference eigenvalue for the mode of order r can be approximated as

$$\lambda_q \simeq \frac{\eta}{2\pi^2 D} \int_{-\infty}^{\infty} \frac{d\beta}{\sqrt{1-\beta^2}} B_q(-\beta) B_q(\beta) \left[1 - (-1)^r e^{-j2\pi W \sqrt{1-\beta^2}} \right] \tag{6.12}$$

where r is an integer to be specified below. The function $B_q(k)$ given in (5.43) has a maximum at $\beta_q = q/(2D)$, and the dominant contribution to the integral comes from a small region about the maximum. In order for the eigenvalue to be small in magnitude, the factor in square brackets must be zero at $\beta = \beta_q$. This leads to the resonance condition

$$W\sqrt{1-\beta_q^2} = \frac{r}{2}, \quad r = \pm 0, \pm 1, \dots \tag{6.13}$$

We note that for the strongly resonant modes $q \neq 0$, since the DC mode does not vanish at the openings of the cavity and therefore radiates to the far field.

The modal fields associated with resonances of the open cavity are approximately given by the nontrivial solutions of the interior Helmholtz problem for a rectangular domain with the Dirichlet boundary condition. This means that the resonances of the parallel strip resonator are close to the internal resonances of a closed rectangular region. By making use of the definition of β_q, it can be seen that (6.13) is equivalent to

$$\sqrt{\left(\frac{q}{2D}\right)^2 + \left(\frac{r}{2W}\right)^2} = 1 \tag{6.14}$$

which is the resonance condition for the TM_{qr} mode of a closed rectangular cavity with side lengths of D and W in wavelengths.

By solving (6.13) for β_q, we find that the normalized spatial frequency of the resonant mode is given by

$$\beta_q = \sqrt{1 - \left(\frac{r}{2W}\right)^2} \tag{6.15}$$

Since β_q is real, the condition $r < 2W$ must hold. As r increases, the angle of propagation of the field radiated by the corresponding mode becomes closer to normal with respect to the cavity walls, and the fields are more strongly confined to the cavity. Thus, the high r resonances radiate little power to the far field and the corresponding eigenvalues have small real parts. In order to find the mode number of the eigenvalue with smallest real part, we choose the largest possible value of r, which is

$$r = \lfloor 2W \rfloor \tag{6.16}$$

where $\lfloor x \rfloor$ is the integer part of x. From (6.15), the corresponding value of the normalized mode spatial frequency is

$$\beta_q \simeq \sqrt{\frac{\alpha}{W}} \tag{6.17}$$

where $\alpha = 2W - \lfloor 2W \rfloor$ is the fractional part of $2W$. With this result, we have identified the resonant mode that will have the smallest eigenvalue and hence will be the most problematic from a numerical point of view for the method of moments.

We now wish to find the operator eigenvalue corresponding to this resonant mode. The real part of the eigenvalue can be estimated by taking the real part of (6.12). After combining the integrand for $\beta > 0$ with $\beta < 0$, we obtain

$$\text{Re}\{\lambda_q\} \simeq \frac{4\eta\beta_q^2}{\pi^2 D} \int_0^1 \frac{dk}{\sqrt{1-\beta^2}} \frac{\sin^2\left[\pi D(\beta - \beta_q)\right]}{(\beta^2 - \beta_q^2)^2} \sin^2\left(\pi W\sqrt{1-\beta^2}\right) \tag{6.18}$$

This integral can be estimated as [4]

$$\text{Re}\{\lambda_q\} \simeq \frac{\eta\sqrt{W}}{2D}\left(\alpha + \frac{\pi^2\alpha^2}{18}\right) \tag{6.19}$$

To leading order, the same result is obtained for the TE-EFIE.

Figure 6.3 compares the real part of the eigenvalue corresponding to the mode closest to resonance as a function of the plate separation distance. We can observe in (6.19) that α, the fractional part of $2W$, vanishes if the width w of the parallel strip resonator is an integer number of half wavelengths. Since the eigenvalue estimate (6.12) is based on the normal operator approximation defined in (1.4), the estimate does not include

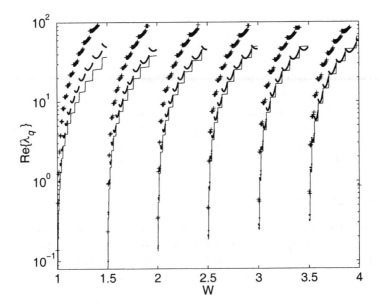

Figure 6.3: Real part of the smallest cavity mode eigenvalue for a parallel strip resonator of depth $L = 10$ as a function of the width. Pluses: computed TM-EFIE operator eigenvalue. Solid line: theoretical estimate, (6.19). Dots: numerical integration of (6.12). Jumps in the real part occur when a new mode moves closest to resonance. (©John Wiley & Sons [4].)

the effects of edge diffraction, and the true operator eigenvalue does not exactly vanish when $2W$ is an integer. Even with the effects of edge diffraction, however, the eigenvalue becomes very small near integer and half-integer cavity widths, as can be seen in Figure 6.3.

As can be seen in Figure 6.3, (6.19) result gives a good estimate for the real part of resonant mode eigenvalues for most values of W. Although it is not readily apparent in the figure, however, the estimate breaks down near integer and half-integer cavity widths, which correspond to the strongest resonances of the structure. If $2W$ is an integer, (6.17) predicts a zero value for β_q, but as observed above the DC mode is not strongly resonant, so we must have $q > 0$. This means that for integer and half-integer widths, the $q = 1$ mode is nearest to resonance. Expanding the resonance condition (6.13) for large D gives the width

$$W_1 = W + \frac{W}{8D^2} \tag{6.20}$$

where W is an integer or half integer. At this value for the cavity width, the $q = 1$ mode is exactly at resonance. This leads to the value $\alpha = W/(4D^2)$, which when substituted into (6.19) leads to

$$\text{Re}\{\lambda_1\} \simeq \frac{\eta W^{3/2}}{8D^3} \qquad (6.21)$$

for the real part of the $q = 1$ resonant mode eigenvalue.

While we have carried through this analysis for the parallel strip resonator, similar results are obtained for the half-open parallel strip resonator. Since the fields radiated by the cavity mode with large r and small β_q propagate predominantly in a direction near normal to the strips, a conducting wall at one end of the resonator has only a small effect on the strongest resonances.

6.3.2 Near-Resonant Case

If the resonance condition (6.13) is not satisfied exactly, then the imaginary part of the smallest cavity mode eigenvalue is nonzero. For a given set of cavity dimensions D and W, let q be the positive integer for which (6.13) is closest to equality, and let W_q be the nearest cavity width for which the resonance condition holds, so that

$$W_q \sqrt{1 - \beta_q^2} = r/2 \qquad (6.22)$$

We then define the shift v_q away from resonance by

$$v_q = W - W_q \qquad (6.23)$$

The imaginary part of the difference eigenvalue as given by (6.12) can be estimated as [4]

$$\text{Im}\{\lambda_q\} \simeq j\eta\pi v_q \qquad (6.24)$$

If the frequency is moved away from a resonance frequency, the eigenvalue maintains a more or less constant real part, but the imaginary part is nonzero. This matches the intuition developed in Section 6.1 with regard to the vertical trajectory followed by the eigenvalue in the complex plane.

To validate these eigenvalue estimates, Figure 6.4 shows the real and imaginary parts of the smallest cavity mode eigenvalue of the TM-EFIE, as a function of the cavity electrical depth D, for an electrical width of $W = 3$. In view of (6.20), the mode approaches resonance as D increases, and the quality factor of the resonance increases as well. As a result, both the real and imaginary parts of the eigenvalue decrease rapidly as the depth of the cavity increases. The theoretical prediction for the real part of the eigenvalue is smaller than the computed value due to the additional power loss caused by edge diffraction.

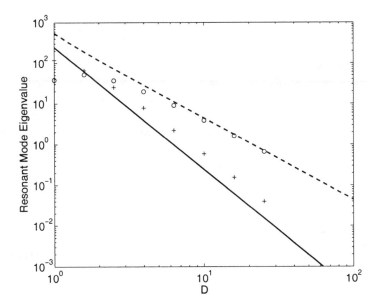

Figure 6.4: Smallest cavity mode eigenvalue as a function of depth for the parallel strip resonator with electrical width $W = 3$. Circles: imaginary part, computed from moment matrix. Pluses: real part, computed from moment matrix. Dashed lines: theoretical estimates, (6.21) and (6.24).

6.3.3 Spectral Error

As noted above, projection error enters into the operator spectrum multiplicatively, so that its relative effect is independent of the magnitude of the eigenvalue. Aliasing error is additive, so that the relative error becomes large as the magnitude of the eigenvalue decreases. Since aliasing error is associated with the singularity of the operator kernel and the kernel is smooth when the source and observation points are well separated, this spectral error contribution affects only the diagonal blocks $\mathbf{L}^s$ in (6.8). For the purposes of this section, we will take the discretization scheme to be point matching with pulse basis functions for the TM-EFIE. Moment matrix integrals will be evaluated with a single-point quadrature rule for off-diagonal matrix elements and a first-order analytical integration of the diagonal elements. From (3.64), the spectral error associated with this discretization scheme is

$$\Delta\lambda_q \simeq -\frac{j\eta}{n_\lambda}\left(\ln\pi - 1\right) \tag{6.25}$$

In the spectral error analysis of previous chapters, the high-order eigenvalues of the EFIE corresponding to nonradiating modes were approximated as purely imaginary.

As a consequence, the aliasing error approximation in (6.25) is purely imaginary. Since the high-order eigenvalues do have a very small real part, the spectral error also has a small real part, which is neglected in (6.25).

Because the spectral error is almost purely imaginary, one effect of discretization is to shift the resonance frequencies of the cavity. As observed earlier in this chapter, eigenvalues of near resonant modes move vertically, parallel to the imaginary axis, with a real part dictated by the finite quality factor of the resonance. The imaginary spectral error shifts the eigenvalues so that they cross the real axis at a different frequency.

This effect can be seen analytically by combining the eigenvalue estimate given by (6.19) and (6.24) with the spectral error (6.25) to obtain

$$\lambda_q \simeq \frac{\eta\sqrt{W}}{2D}\left(\alpha + \frac{\pi^2\alpha^2}{18}\right) + j\left[\eta\pi\nu_q - \frac{\eta}{n_\lambda}\left(\ln\pi - 1\right)\right] \qquad (6.26)$$

In view of (6.23), the resulting shift in the location of the resonance considered as a function of the electrical cavity width W is approximately

$$\Delta W \simeq \frac{0.05}{n_\lambda} \qquad (6.27)$$

The relative resonance frequency shift is

$$\frac{\Delta k}{k_0} \simeq \frac{0.05}{n_\lambda W} \qquad (6.28)$$

From this result, it can be seen that the shift in the location of the resonance is small if the width of the resonator is larger than the wavelength. For narrow resonators ($W < 1$), however, the shift in the resonance can be very large. This accounts for the numerical difficulties with the method of moments for scatterers with closely spaced conducting surfaces.

To validate the spectral estimate (6.26), Figure 6.5 compares the imaginary part of the smallest resonant eigenvalue of the moment matrix for open and half-open parallel strip resonators for $W = 2$ and $n_\lambda = 10$ to the theoretical estimates obtained above, in the presence of discretization error. A single-point integration rule with analytical self-interactions is used to compute moment matrix elements. For $W = 2$, the lowest-order cavity mode is nearest to resonance. For finite D, the mode is not exactly at resonance, so the exact eigenvalue has a nonzero imaginary part, but as D increases, the mode moves closer to resonance and the imaginary part tends to zero. When the EFIE is discretized, the imaginary part of the eigenvalue tends to the value of the spectral error (6.25) as the exact eigenvalue becomes small.

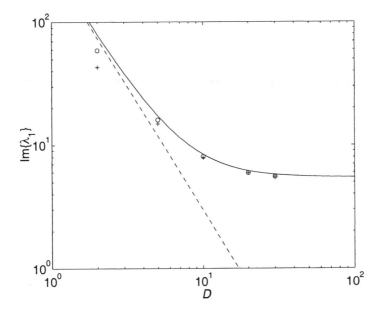

Figure 6.5: Imaginary part of smallest resonant eigenvalue for a parallel strip resonator as a function of depth with the width fixed at $W = 2$. Integral equation: EFIE. Polarization: TM. Mesh density: $n_\lambda = 10$. Pluses: computed moment matrix eigenvalue, open cavity. Circles: computed moment matrix eigenvalue, half-open cavity. Solid line: theoretical eigenvalue estimate, including discretization error (6.25). Dashed line: theoretical result, without discretization error. (©John Wiley & Sons [4].)

6.3.4 Scattering Amplitude Error

From (5.36), the relative scattering amplitude error neglecting edge diffraction is given by the aliasing component of the relative spectral error. That analysis was given for a single flat strip, but the same error estimate can be applied to the parallel strip resonator. For the discretization considered in the previous section, the spectral error is first order in n_λ^{-1}. Since the error contribution from edge diffraction given by (5.56) is also first order and is not dominant for this discretization scheme, (5.36) can be used as an estimate for the total scattering amplitude error. This reasoning leads to the estimate

$$\frac{|S - \hat{S}|}{|S|} \simeq \frac{\eta}{n_\lambda |\lambda_q|} (\ln \pi - 1) \tag{6.29}$$

From the TM-EFIE spectrum shown in Figure 5.1, it can be seen that λ_q for non-resonant modes is of order $\eta/2$. With this eigenvalue estimate, the relative scattering

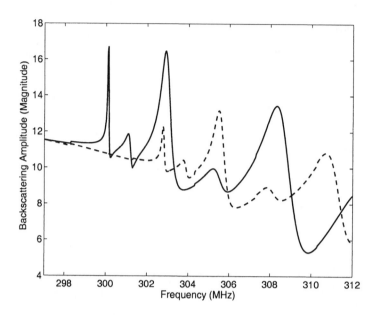

Figure 6.6: Magnitude of backscattering amplitude for a parallel strip resonator, $D = 10$, $W = 0.5$. Integral equation: EFIE. Polarization: TM. Incidence angle: 67.5°. Solid line: reference solution. Dashed line: MoM, single-point integration rule with analytical diagonal moment matrix elements, $n_\lambda = 10$.

amplitude error away from resonances is

$$\frac{|S - \hat{S}|}{|S|} \simeq \frac{0.3}{n_\lambda} \tag{6.30}$$

Near a resonance frequency of the structure, the magnitude of the eigenvalue of the resonant mode is much smaller than $\eta/2$. Consequently, the relative spectral error (6.29) is larger and the scattering amplitude error increases significantly.

Figure 6.6 shows the magnitude of the backscattering amplitude of a parallel strip resonator. A numerical solution using MoM with the single-point integration rule considered above is compared to a reference solution obtained with a high accuracy discretization of the TM-EFIE. At 297 MHz, the scatterer is not resonant, and the scattering amplitude error is small. For a mesh density of $n_\lambda = 10$, the relative scattering amplitude error estimate (6.30) is 3%. This error is small enough that it cannot be read from Figure 6.6. The numerical error observed at 297 MHz is 3.2%, which is very close to the theoretical estimate. The increase in error near resonance frequencies is readily apparent in Figure 6.6. At the resonant peak near 300 MHz, for example, the relative scattering amplitude error is 40%. It is also apparent in Figure 6.6 that the resonant frequencies of the structure are shifted by discretization. The pattern of resonant peaks for the MoM results is shifted by a few MHz relative to the reference solution.

6.4 SUMMARY

We have examined the behavior of the method of moments for closed scatterers with internal resonances and cavities with physical resonances coupled to external fields. The resonance properties of a structure are reflected by the integral operator spectrum. The real part of the eigenvalue corresponding to a resonant mode represents energy radiated by the mode to the far field. The smaller the real part, the higher the quality factor of the resonance. The imaginary part is the difference between the energies stored by the scattered magnetic and electric fields associated with the mode. At resonance, the magnetic and electric energies are in balance and the eigenvalue is purely real. For an internal resonance, the real part is zero and the eigenvalue vanishes. For both types of resonance, the magnitude of the eigenvalue is small, which magnifies the relative impact of the additive spectral error caused by discretization. This leads to large solution error near resonance frequencies. Furthermore, since the additive part of the spectral error is nearly pure imaginary, discretization shifts the resonance frequencies of the structure.

In order to improve accuracy of the moment method for resonant structures, a discretization scheme with small aliasing error is required. From the results of earlier chapters, this can be achieved by using smooth basis functions and highly accurate moment matrix integration rules.

REFERENCES

[1] J. A. Kong, *Electromagnetic Wave Theory*. New York: John Wiley & Sons, 2nd ed., 1990.

[2] J. M. Song, W. W. Shu, and W. C. Chew, "Numerical resonances in method of moments," *Proceedings of the IEEE Antennas and Propagation Society International Symposium*, Honolulu, HI, June 2007.

[3] C. P. Davis, "Understanding and improving moment method scattering solutions," Master's thesis, Brigham Young University, 2004.

[4] K. F. Warnick and W. C. Chew, "Convergence of moment method solutions of the electric field integral equation for a 2D open cavity," *Microw. Opt. Tech. Lett.*, vol. 23, pp. 212–218, Nov. 1999.

Chapter 7

Error Analysis for 3D Problems

The two-dimensional scattering problems we have considered up to this point provide a mathematically simple framework that is ideal for developing key concepts of numerical analysis. Since real-world applications deal primarily with three-dimensional structures, it is important to extend these techniques and results to the 3D case.

From a computational point of view, the main challenges with 3D problems are the complexity of implementing vector integral equations and vector basis functions for curved meshes and the large number of degrees of freedom required for 3D problems. Over the past decade, the computational electromagnetics community has largely worked through the implementation details for vector integral equations, and commercial software packages for 3D structures are widely available. Dealing with the large number of degrees of freedom for 3D problems still represents a significant computational challenge. If ten unknowns per wavelength is adequate for a 2D scatterer, a similar 3D problem may require 100 unknowns per square wavelength, which places a severe limit on the electrical size of problems that can be solved numerically, although advances in fast algorithms and parallelization have greatly expanded the realm of computationally feasible radiation and scattering problems.

Understanding of solution error behavior for 3D problems has lagged behind algorithm development. To provide solution error estimates and insight for 3D problems, we consider the flat PEC plate as a canonical scatterer and obtain spectral error estimates for the method of moments for the vector EFIE. One of the goals will be to understand the connections between the 2D analyses of earlier chapters and 3D problems. From an implementation point of view, 3D algorithms can be much more complex than 2D, but from a physical point of view, there are close connections between the phenomena associated with 2D scattering for the TM and TE polarizations and 3D scattering. For example, we will find that the EFIE operator spectrum for the flat plate is essentially a combination of the spectra for the 2D TM-EFIE and TE-EFIE operators for the flat strip. The benefit of understanding these connections is that the results and insight developed for 2D problems can be applied to 3D numerical methods.

7.1 FLAT PLATE

In Section 5.1, we observed that spectral estimates can be obtained for a flat strip by approximating the eigenfunctions of the two-dimensional EFIE as modes of the form $e^{jk_0\beta x}$. A similar modal expansion can be used for a flat PEC plate with dimensions $d \times d$. The EFIE operator $\mathcal{T}$ is nonnormal, and the Fourier modes are not exact eigenfunctions of the operator, but we can apply the normal decomposition (1.4) to obtain eigenvalue estimates. These eigenvalue estimates can be obtained for both the exact integral operator and the moment matrix discretization, which leads to spectral error and solution error estimates for the method of moments.

We will approximate the vector eigenfunctions of the EFIE operator as

$$\mathbf{v} = \hat{t} e^{jk_0\boldsymbol{\beta}\cdot\mathbf{r}} \tag{7.1}$$

where $\hat{t}$ is a constant unit vector tangential to the scatterer surface and

$$\boldsymbol{\beta} = \beta_x \hat{x} + \beta_y \hat{y} \tag{7.2}$$

defines the normalized spatial frequency of the mode. Applying the operator to this mode and making use of the 2D Fourier representation of the scalar Green's function yields

$$\mathcal{T}\hat{t}e^{jk_0\boldsymbol{\beta}\cdot\mathbf{r}} = \frac{k_0\eta d^2}{8\pi^2} \int dk_x\, dk_y \frac{e^{j\mathbf{k}\cdot\mathbf{r}}}{k_z} \frac{\sin\left[(k_x - k_0\beta_x)d/2\right]}{(k_x - k_0\beta_x)d/2} \frac{\sin\left[(k_y - k_0\beta_y)d/2\right]}{(k_y - k_0\beta_y)d/2}$$
$$\times \left[\hat{t} - \frac{\mathbf{k}(\boldsymbol{\beta}\cdot\hat{t})}{k_0}\right] \tag{7.3}$$

where $k_z = \sqrt{k_0^2 - k_x^2 - k_y^2}$. By expanding the integrand about the maxima of the sinc functions at $\mathbf{k} = k_0\boldsymbol{\beta}$, (7.3) can be approximated as

$$\mathcal{T}\hat{t}e^{jk_0\boldsymbol{\beta}\cdot\mathbf{r}} \sim \frac{\eta d^2}{8\pi^2} \frac{e^{jk_0\boldsymbol{\beta}\cdot\mathbf{r}}}{\sqrt{1-\beta^2}} \left[\hat{t} - \boldsymbol{\beta}(\boldsymbol{\beta}\cdot\hat{t})\right] \int dk_x\, dk_y \tag{7.4}$$

$$\times \frac{\sin\left[(k_x - k_0\beta_x)d/2\right]}{(k_x - k_0\beta_x)d/2} \frac{\sin\left[(k_y - k_0\beta_y)d/2\right]}{(k_y - k_0\beta_y)d/2}, \quad d \to \infty$$

Evaluating the integrals yields

$$\mathcal{T}\hat{t}e^{jk_0\boldsymbol{\beta}\cdot\mathbf{r}} \sim \frac{\eta}{2\sqrt{1-\beta^2}} \left[\hat{t} - \boldsymbol{\beta}(\boldsymbol{\beta}\cdot\hat{t})\right] e^{jk_0\boldsymbol{\beta}\cdot\mathbf{r}}, \quad d \to \infty \tag{7.5}$$

Equation (7.5) indicates that the eigenvalues of the EFIE operator can be divided into two groups. If $\hat{t}$ is parallel to β, then the vector in square brackets is a scalar multiple of $\hat{t}$, and $\hat{t}e^{jk_0\beta \cdot \mathbf{r}}$ is an approximate eigenfunction of $\mathcal{T}$, with the eigenvalue

$$\lambda^{\text{TE}} \simeq \frac{\eta}{2}\sqrt{1-\beta^2} \tag{7.6}$$

This result corresponds to the curl-free modes on the scatterer, which are essentially the same as the eigenfunctions of the 2D TE-EFIE operator $\mathcal{N}$. For the divergence-free modes, $\hat{t}$ is perpendicular to β, and (7.5) yields the approximate eigenvalue

$$\lambda^{\text{TM}} \simeq \frac{\eta}{2}\frac{1}{\sqrt{1-\beta^2}} \tag{7.7}$$

These modes are similar to eigenfunctions of the 2D TM-EFIE operator $\mathcal{L}$.

Because the expansion used to obtain (7.5) fails near the singularity of the integrand in (7.3), these eigenvalue estimates break down as β approaches the singularities at $|\beta| = 1$. These correspond to the surface wave current modes with spatial frequency near k_0. Equation (7.6) predicts a zero value for the eigenvalue of the curl-free surface wave mode, and (7.7) becomes infinite for the divergence-free surface wave mode, whereas the actual operator eigenvalues are finite and nonzero. It would be possible to develop eigenvalue estimates for the surface wave modes, but for the purpose of error analysis, (7.6) and (7.7) are adequate.

The operator $\mathcal{T}$ has an infinite number of eigenvalues, with accumulation points at the origin and at $-j\infty$. The largest and smallest eigenvalues correspond to nonradiating modes with large spatial frequency β. The curl-free or TE type nonpropagating modes have large eigenvalues, due to the divergence operator in (2.5). The divergence-free or TM type nonpropagating modes have small eigenvalues. These behaviors for nonpropagating modes are identical to those observed in earlier chapters for the 2D TE-EFIE and TM-EFIE operators.

7.1.1 Moment Matrix Spectrum

When the EFIE is discretized, the resulting matrix operator has a finite number of eigenvalues, corresponding to the range of spatial frequencies representable on the mesh. The highest representable spatial frequency or mesh Nyquist frequency (2.63) corresponds to a maximum normalized spatial frequency of

$$\beta_{\max} = \frac{n_\lambda}{2} \tag{7.8}$$

The eigenvectors of the moment matrix correspond approximately to the eigenfunctions of $\mathcal{T}$ with normalized spatial frequency β in the range $-n_\lambda/2 \leq \beta \leq n_\lambda/2$. For a two-dimensional mesh, mesh elements typically have different edge lengths, so there is not a unique way to define the mesh length h used to determine the linear mesh density $n_\lambda = \lambda/h$. Some common choices are the average element edge length or the maximum edge length.

The largest eigenvalue of the moment matrix corresponds to a curl-free or TE type mode with spatial frequency near the mesh Nyquist frequency. Using (7.8) in (7.6) leads to the maximum moment matrix eigenvalue estimate

$$\lambda_{\max}^{\text{TE}} \simeq -\frac{j\eta n_\lambda}{4} \tag{7.9}$$

The smallest eigenvalue is associated with a divergence-free mode and can be obtained from (7.7) as

$$\lambda_{\min}^{\text{TM}} \simeq \frac{j\eta}{n_\lambda} \tag{7.10}$$

In Chapter 9, these eigenvalue estimates will be used to analyze the conditioning of the moment matrix, which is a measure of the difficulty of solving a linear system for the MoM current solution. These estimates are valid for low-order basis functions. For higher-order basis functions (p-refinement), a more sophisticated treatment will be developed in Chapter 8.

The moment matrix spectrum for a flat PEC plate is shown in Figure 7.1. By comparison with Figures 5.1 and 5.4, it can be seen that the spectrum for the 3D EFIE operator can be viewed as a combination of the eigenvalues of the TM-EFIE eigenvalues and the TE-EFIE eigenvalues for the flat strip. The eigenvalues lie on a question mark shaped arc in the complex plane that crosses the real axis at roughly $(\eta/2, 0)$, which is the eigenvalue of the DC (zero spatial frequency) current mode for both the TM-MFIE and TE-EFIE.

To validate the spectral estimates obtained above, we will compare the numerical spectrum shown in Figure 7.1 to the theoretical eigenvalue estimates. For a mesh element density of $n_\lambda = 10$, $\lambda_{\max}^{\text{TE}} \simeq -j940\ \Omega$, which is of the same order as the eigenvalue with largest magnitude observed in Figure 7.1. For $n_\lambda = 10$, we obtain $\lambda_{\min}^{\text{TM}} \simeq j38\ \Omega$. Numerically, the smallest moment matrix eigenvalue is $0.2 + j37\ \Omega$, which is very close to the theoretical estimate. In order to analyze the moment method solution error for the EFIE, we will now consider the spectral error introduced by discretization for a particular set of basis functions.

7.1.2 Rooftop Basis Functions

For the purposes of spectral error analysis, the rooftop basis functions are convenient, because rooftop functions are more homogeneous across a rectangular mesh than basis functions defined on triangular patches. The rooftop basis is a simple 2D generaliza-

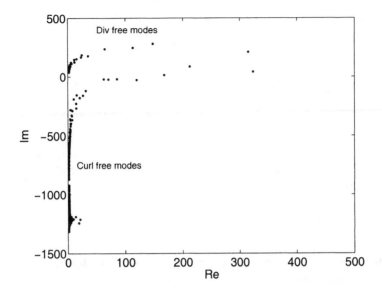

Figure 7.1: Moment matrix spectrum for 1m × 1m flat PEC plate at 300 MHz. Integral equation: EFIE. Mesh: 200 triangular patches, 280 edges, 10 elements per wavelength. Basis: RWG. The spectrum is roughly a combination of the TM-EFIE eigenvalues (corresponding to divergence-free modes) for the flat strip shown in Figure 5.1 and the TE-EFIE eigenvalues (curl-free modes) in Figure 5.4.

tion of the 1D pulse and triangle functions. The triangle portion of the basis function allows the divergence operators in the EFIE operator $\mathcal{T}$ to be evaluated in closed form.

We define two sets of node points,

$$\mathbf{r}_{mn1} = mh\hat{x} + (n - 1/2)h\hat{y}, \quad m = 1, \ldots, M - 1, \quad n = 1, \ldots, M \qquad (7.11a)$$

$$\mathbf{r}_{mn2} = (m - 1/2)h\hat{x} + nh\hat{y}, \quad m = 1, \ldots, M, \quad n = 1, \ldots, M - 1. \qquad (7.11b)$$

For convenience, we assume that $d = Mh$, so that the plate is an even number of discretization lengths in size. The rooftop functions are

$$\mathbf{f}_{mnp}(\mathbf{r}) = \mathbf{f}_p(\mathbf{r} - \mathbf{r}_{mnp}), \quad p = 1, 2 \qquad (7.12)$$

where

$$\mathbf{f}_1(\mathbf{r}) = \begin{cases} \hat{x}(1 - |x|/h) & -h \leq x \leq h, \ -h/2 \leq y \leq h/2 \\ 0 & \text{otherwise} \end{cases} \qquad (7.13)$$

$$\mathbf{f}_2(\mathbf{r}) = \begin{cases} \hat{y}(1 - |y|/h) & -h/2 \leq x \leq h/2, \ -h \leq y \leq h \\ 0 & \text{otherwise} \end{cases} \qquad (7.14)$$

These basis functions have a linear dependence in one direction of the same form as the triangle function, and are constant in the orthogonal direction.

For the rooftop basis, the moment matrix elements are

$$Z_{mnp,m'n'p'} = jk_0\eta h^{-2} \int \int d\mathbf{r} \, d\mathbf{r}' \, g(\mathbf{r},\mathbf{r}') \Big[\mathbf{f}_{mnp}(\mathbf{r}) \cdot \mathbf{f}_{m'n'p'}(\mathbf{r}')$$
$$+ k_0^{-2} \nabla \cdot \mathbf{f}_{mnp}(\mathbf{r}) \nabla' \cdot \mathbf{f}_{m'n'p'}(\mathbf{r}') \Big] \tag{7.15}$$

We will obtain moment matrix spectral estimates using the approximate eigenfunctions $\hat{t}e^{-jk_0\beta\cdot\mathbf{r}}$ evaluated at the node points $\mathbf{r}_{mnp}$. We transform the moment matrix according to

$$\hat{\lambda} \simeq \frac{1}{M^2} \sum_{mnp,m'n'p'} t_p e^{-jk_0\beta\cdot(\mathbf{r}_{mnp}-\mathbf{r}_{m'n'p'})} Z_{mnp,m'n'p'} \tag{7.16}$$

where t_p, $p = 1,2$ indexes the x and y components of $\hat{t}$. By adapting the derivation leading to (7.3), we obtain

$$\hat{\lambda} \simeq \frac{k_0\eta h^2}{8M^2\pi^2} \int \frac{d\mathbf{k}}{k_z} \sum_{p,p'} t_p t_{p'} R^2(\mathbf{k}-k_0\beta)$$
$$\times \left[\mathbf{F}_p(-\mathbf{k}) \cdot \mathbf{F}_{p'}(\mathbf{k}) - \frac{1}{k_0^2} \mathbf{k} \cdot \mathbf{F}_p(-\mathbf{k}) \, \mathbf{k} \cdot \mathbf{F}_{p'}(\mathbf{k}) \right] \tag{7.17}$$

where

$$R(\mathbf{k}) = \frac{\sin(k_x d/2)}{\sin(k_x h/2)} \frac{\sin(k_y d/2)}{\sin(k_y h/2)} \tag{7.18}$$

$$\mathbf{F}_1(\mathbf{k}) = \hat{x} \left[\frac{\sin(k_x h/2)}{k_x h/2} \right]^2 \frac{\sin(k_y h/2)}{k_y h/2} \tag{7.19}$$

$$\mathbf{F}_2(\mathbf{k}) = \hat{y} \frac{\sin(k_x h/2)}{k_x h/2} \left[\frac{\sin(k_y h/2)}{k_y h/2} \right]^2 \tag{7.20}$$

By expanding the integrand about each of the maxima of $R^2(\mathbf{k}-k_0\beta)$ as in the derivation of (5.15), we arrive at the estimate

$$\hat{\lambda} \simeq \frac{k_0\eta}{2} \sum_{q,r=-\infty}^{\infty} \sum_{p,p'=1}^{2} \frac{1}{k_{qrz}} t_p t_{p'} F_{pp}(\mathbf{k}_{qr}) F_{p'p'}(\mathbf{k}_{qr})(\delta_{pp'} - k_{qrp}k_{qrp'}/k_0^2) \tag{7.21}$$

where

$$\mathbf{k}_{qr} = \hat{x}(k_0\beta_x + 2\pi q/h) + \hat{y}(k_0\beta_y + 2\pi r/h) \tag{7.22}$$

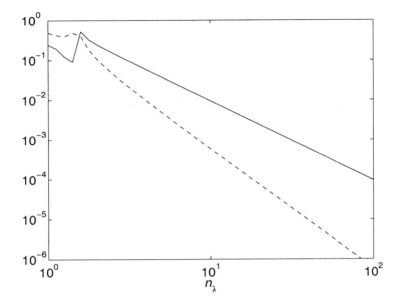

Figure 7.2: Spectral error for the EFIE with rooftop basis functions as a function of discretization density, for a divergence-free mode with normalized spatial frequency $\beta = 1/2$. Solid line: total error, (7.23). Dashed line: aliasing error contribution.

The second subscript on F_{pp} indexes the x and y components of $\mathbf{F}_p$, and the subscript p on k_{qrp} denotes the x and y components k_{qrx} and k_{qry}.

From the definition (3.16), the spectral error is

$$\Delta\lambda \simeq \frac{k_0}{2\sqrt{1-\beta^2}} \sum_{p,p'=1}^{2} t_p t_{p'} \left[F_{pp}(k_0\beta)F_{p'p'}(k_0\beta) - 1 \right] (\delta_{pp'} - \beta_p\beta_{p'})$$

$$+ \frac{k_0\eta}{2} \sum_{q,r\neq 0} \sum_{p,p'=1}^{2} \frac{1}{k_{qrz}} t_p t_{p'} F_{pp}(\mathbf{k}_{qr}) F_{p'p'}(\mathbf{k}_{qr}) (\delta_{pp'} - k_{qrp}k_{qrp'}/k_0^2) \qquad (7.23)$$

This expression is approximate because the spectral error is obtained from eigenvalue estimates rather than exact expressions. As with the 2D flat strip, for which exact eigenvalues were not available, we are more interested in the spectral error caused by discretization than in the eigenvalue itself, and the use of estimates suffices for this purpose.

The spectral error (7.23) can be divided into projection and aliasing error contributions. Projection error is associated with the $q = r = 0$ term of the summation in (7.21), and the remaining terms represent the aliasing error. We can determine the or-

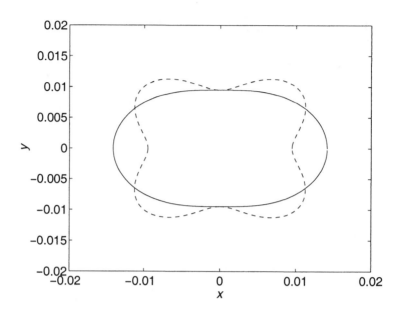

Figure 7.3: Polar representation of spectral error as a function of mode current flow and wave vector directions for a fixed mesh density ($n_\lambda = 10$). Solid line: $\hat{t}$ rotated; β fixed at $\hat{x}/2$. Dashed line: both $\hat{t}$ and β rotated.

der of the spectral error in h by inspection of (7.23). The projection error is of order h^2, which can be seen by expanding $R(\mathbf{k})$ for small k. For the aliasing error, the term $1/k_{qrz}$ contributes a factor of h, the term $F_{pp}F_{p'p'}$ contributes a factor of at most h^4, and $k_{qrp}k_{qrp'}$ contributes a factor of h^{-2}, for an overall order of h^3, or n_λ^{-3}.

Figure 7.2 shows the spectral error as a function of the mesh density. It can be seen that the orders of the projection and aliasing errors are second and third order, as predicted. For these results, $\hat{t} = \hat{x}$ and $\beta = \hat{y}/2$. Since the direction and phase vectors are perpendicular, the mode for which the spectral error is shown is a divergence-free or TM-type mode.

For modes on a 2D mesh, spectral error depends on the current flow direction $\hat{t}$ as well as the direction of spatial variation given by the normalized wave vector $\hat{\beta}$. While the convergence order of the spectral error is the same for any direction, the magnitude of the error varies with $\hat{t}$ and $\hat{\beta}$ in a way that is similar to the dependence of dispersion error on the wave propagation direction for the finite difference time domain algorithm (FDTD). This is illustrated in Figure 7.3, which shows the spectral error for a fixed mesh density for different mode current flow and wave vector directions. The spectral error is largest when $\hat{t}$ and β are parallel. In this case, the mode has the largest divergence, and the error contribution from the hypersingular term of the EFIE is greatest. Thus, the curl-free or TE type modes have larger spectral error than TM type modes. Error is smallest for propagation directions that are vertical or horizontal with respect to the rectangular grid used to define the rooftop basis functions, as can be seen from the dashed curve in Figure 7.3.

From the analysis of this section, we can see that the convergence rate of the spectral error is identical to that of the low-order bases studied in Sections 3.1.3, 3.4.1, and 5.1.3 for the 2D case with ideal discretizations. This indicates that the numerical behavior of the method of moments for the 3D EFIE is closely related to that observed with the 2D TM-EFIE and TE-EFIE operators.

This error analysis ignores current singularities at the scatterer edges and is valid only for the interior part of the flat plate. The predicted convergence rates should match those observed for smooth scatterers without edge or corner singularities. We have seen in Chapter 5 that spectral error and solution error estimates for the smooth circular cylinder and other smooth scatterers are similar to those for the interior of the flat strip, ignoring edge singularities. By extension, the convergence order estimates for the interior of a flat plate can be expected to hold for other smooth scatterers, such as the PEC sphere for which results are given in the next section.

7.2 RWG Basis Functions

Because the Rao-Wilton-Glisson (RWG) basis functions [1, 2] are defined on a triangular mesh, an analytical treatment along the lines of the previous section for rooftop basis functions is tedious. Fortunately, since the rooftop and RWG basis functions have similar properties, such as finite divergence and a combination of constant and linear variations, the spectral error for RWG basis functions can be expected to be close to that obtained in the previous section for the rooftop basis, with a second-order projection error component and third-order aliasing contribution.

In view of this relationship between rooftop and RWG functions, the spectral error analysis of the previous section provides predicted current and solution error convergence rates which can be compared to numerical results with the RWG basis. Figure 7.4 shows the current solution error for a PEC sphere. RWG functions are used for both testing and expansion (Galerkin's method). The error is shown as a function of linear mesh element density, which here is defined to be $n_\lambda = \lambda/h_{max}$, where h_{max} is the longest element edge length. For the smooth 2D scatterers studied in earlier chapters, the RMS current solution error has the same order as the projection error component of the spectral error, when the error is measured at the mesh element centers. Since the projection error is second order, we should also expect a second-order current error. This requires that the error be measured at discrete points where the basis functions have an interpolatory property analogous to the one-dimensional basis functions

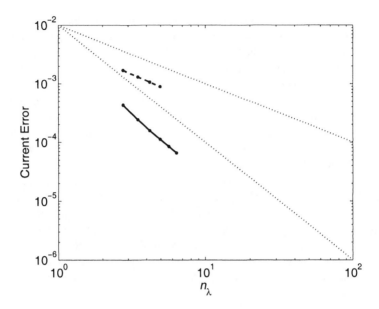

Figure 7.4: Current solution error for a PEC sphere with radius 0.5 m at 300 MHz. Integral equation: EFIE. Basis functions: RWG testing and expansion (Galerkin's method). Discretization: ideal, high accuracy numerical integration and curved mesh elements. Solid line: RMS error at mesh element edge centers. Dashed line: RMS error at edge endpoints. The dotted lines have slopes of −1 and −2. (Results courtesy of A. F. Peterson.)

studied in earlier chapters. For RWG basis functions, the current is most accurate at the mesh element edge centers, and it can be seen in Figure 7.4 that the RMS error is second order when measured at edge centers. Interpolation error causes first-order solution convergence at other points as discussed in Section 2.8. For the RWG basis functions, this leads to a first-order current error when measured at edge endpoints.

Figure 7.5 shows the RCS error for the same scatterer, normalized to the peak RCS of $9.25\,\lambda^2$. The error measure is the RMS copolarized bistatic RCS error over five-degree increments in θ and ϕ and weighted with a $\sin\theta$ factor to compensate for the extra points near the poles of the observation sphere. The RCS error is third order in the linear mesh element density. This is identical to the third-order convergence rate observed for scattered fields in earlier chapters for the 2D TM-EFIE and TE-EFIE with low-order basis function on smooth scatterers and an ideal implementation of the moment method.

Along the lines of the discussion in Section 4.5.1.1, we can arrive at a prediction for the third-order RCS error convergence rate observed in Figure 7.5 by inspection of the

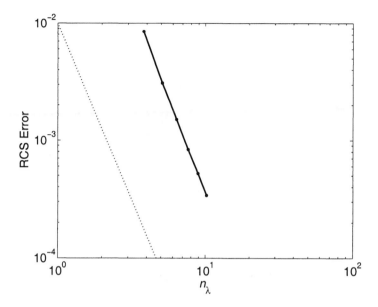

Figure 7.5: Relative RMS RCS error for the scatterer of Figure 7.4. The dotted line has a slope of −3. (Results courtesy of A. F. Peterson.)

smoothness orders of factors in the moment matrix element expression

$$Z_{mn} = \frac{jk_0\eta}{h^2} \iint d\mathbf{r}\,d\mathbf{r}' \underbrace{g(\mathbf{r},\mathbf{r}')}_{O(k^{-1})} \left[\underbrace{\mathbf{f}_m(\mathbf{r})}_{O(k^{-1})} \cdot \underbrace{\mathbf{f}_n(\mathbf{r}')}_{O(k^{-1})} + k_0^{-2} \underbrace{\nabla \cdot \mathbf{f}_m(\mathbf{r})}_{O(k^{-1})} \underbrace{\nabla' \cdot \mathbf{f}_n(\mathbf{r}')}_{O(k^{-1})} \right] \quad (7.24)$$

With the first term in square brackets, the RWG basis functions combine both constant and linear shape functions. The asymptotic falloff of the Fourier transforms of the basis functions is dominated by the constant or pulse-type dependence, so the Fourier transforms decay as k^{-1}. The divergence operators act on the testing and expansion RWG functions, yielding a constant value on each mesh element. This is analogous to a pulse function with a triangular region of support. The Fourier transforms of these constant functions are asymptotically $O(k^{-1})$. The 2D Fourier transform of the scalar Green's function also has order $O(k^{-1})$. The product of the Fourier transforms has order $O(k^{-3})$. Since the eigenfunctions of the EFIE operator for a smooth scatterer are approximately Fourier modes, this is essentially an asymptotic eigenvalue estimate. By comparing this analysis to the results of Chapter 3, it is apparent that the RWG basis is similar to discretizing the TM-EFIE with pulse functions and the TE-EFIE with triangle functions.

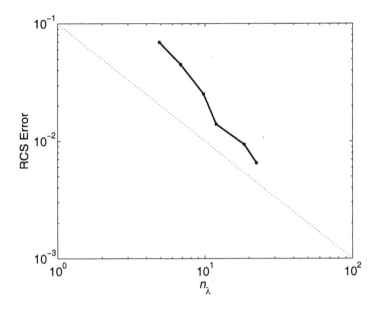

Figure 7.6: Relative RMS RCS error for a PEC sphere with radius 1 m at 300 MHz. Integral equation: EFIE. Basis functions: RWG testing and expansion functions. Discretization: nonideal, with single-point evaluation of moment matrix integrals and flat mesh elements. The dotted line has a slope of −1.

The order of the aliasing error can be estimated from the spectral error contribution of the first aliased mode, or the $q = r = 1$ term of the series in (7.21). Equation (7.21) was derived for the rooftop basis, but a similar (albeit more complex) expression holds for the RWG basis. In the $q = r = 1$ term, an eigenvalue estimate is evaluated for a mode with spatial frequency $k = 2k_{max} = k_0 n_\lambda$. Since the eigenvalue is $O(k^{-3})$ with respect to the spatial frequency, the aliasing error has order n_λ^{-3} or h^3. Because the aliasing error determines the scattering amplitude and RCS errors for an ideal discretization, this simple convergence order argument predicts the third-order convergence rate observed in Figure 7.5.

It should be noted that these arguments hold only for discretizations that are regular in the sense of Section 4.5.1, for which there are no cancellations in the spectral error summation that lead to higher than expected accuracy. These estimates also ignore edge error, which must be analyzed separately using the methods of Section 5.3.

7.2.1 Nonideal Discretizations

The results in the previous section were for an ideal implementation, with accurate numerical integration of moment matrix elements and curved mesh elements. For a

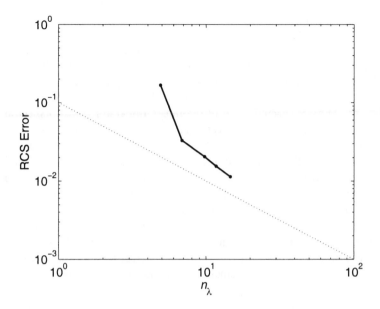

Figure 7.7: Relative RMS RCS error for a PEC sphere with radius 1 m at 300 MHz. Integral equation: MFIE. Basis functions: RWG testing and expansion functions. Discretization: nonideal, with single-point evaluation of moment matrix integrals and flat mesh elements. The dotted line has a slope of −1.

nonideal implementation of the moment method, quadrature error and geometrical discretization error reduce the solution accuracy. Figure 7.6 shows RCS error for a PEC sphere with 1λ radius. The error measure is RMS error for the copolarized bistatic RCS for a θ cut relative to the maximum RCS. The sphere is illuminated by a plane wave incident from $\theta = 0$. Flat mesh elements with a single-point quadrature rule for evaluation of moment matrix elements are employed. Self-interaction terms are evaluated with four integration points.

The RCS convergence rate shown in Figure 7.6 is close to the first-order rate predicted by a simple analysis of (7.24). With a low-order quadrature rule, the basis functions are effectively delta functions with Fourier transforms of order $O(1)$, rather than $O(k^{-1})$, for an overall order of the Fourier transforms of the factors in the matrix element contribution of $O(k^{-1})$. This corresponds to an aliasing error convergence rate of n_λ^{-1}. If moment matrix element integrations are evaluated more accurately so that geometrical discretization error due to flat-facet mesh elements is the dominant source of error, solution accuracy improves to second order.

Figure 7.7 shows numerical results for the MFIE. The first-order convergence rate matches that obtained theoretically in Section 4.2.3 for the 2D MFIE with geometrical discretization error.

7.3 SUMMARY

In this chapter, we have seen close parallels between the error behavior of the method of moments for 3D scattering problems and the 2D behaviors studied in earlier chapters. While the spectral error analysis is mathematically more complex for 3D vector integral operators, the governing principles of error behavior are the same as for the simpler 2D case. Discretization of the integral operator leads to spectral error, which determines the current and scattering amplitude solution errors. For an ideal implementation with low-order rooftop or RWG basis functions, the error for scattered fields with the EFIE was third order, matching results obtained in earlier chapters for 2D scatterers. Convergence orders can be obtained from a simple analysis of the smoothness properties of the operator kernel and basis functions, as long as the discretization is regular in the sense of Section 4.5.1.

Nonideal discretizations cause a decrease in accuracy and a reduction in the solution convergence order, due to quadrature error and geometrical discretization error. At ten unknowns per wavelength, the relative RCS error for an ideal discretization of the EFIE is 0.03%, which is very small. For a nonideal discretization, the RCS error increases dramatically to roughly 2%.

REFERENCES

[1] S. M. Rao, D. R. Wilton, and A. W. Glisson, "Electromagnetic scattering by surfaces of arbitrary shape," *IEEE Trans. Ant. Propag.*, vol. 30, pp. 409–418, May 1982.

[2] A. F. Peterson, R. Mittra, and S. L. Ray, *Computational Methods for Electromagnetics*. New York: IEEE Press, 1998.

Chapter 8

Higher-Order Basis Functions

with Andrew F. Peterson

Thus far we have considered only low-order basis functions—delta functions, pulse functions, and triangle functions, and their two-dimensional vector generalizations, rooftop and RWG functions. While these types of basis functions are commonly used in practice and an understanding of their numerical behaviors is important, higher-order basis functions offer a significant gain in computational efficiency and it would be desirable to understand in detail their impact on solution accuracy.

Higher-order polynomial basis functions can be indexed by the polynomial order p of the highest-order polynomial function included in the basis set. Increasing the polynomial order of a basis without increasing the mesh element density is referred to as p-refinement. The spline basis functions analyzed in earlier chapters can be extended to polynomial orders beyond $p = 0$ and $p = 1$, but those spline type basis functions are not complete, since only one smooth polynomial is employed to represent the current solution per element. Here, we will consider complete polynomial bases, so that on each element the current solution is a polynomial of order p.

Since a polynomial space of order p can be spanned by different elementary polynomials, there are various choices for an order p basis. Some possible choices are monomials $(1, x, \ldots, x^p)$, Lagrange polynomials, and orthogonal polynomials such as the Legendre polynomials. Monomials typically are not used due to numerical instabilities associated with the nonuniformity of the functions across an element at high orders. For vector basis functions, the range of possibilities further increases, since the polynomial order of each component can be chosen independently.

The goal with higher-order bases is to represent a smoothly varying function with as few as possible degrees of freedom, thereby reducing the size of the linear system that must be solved to obtain a numerical solution and improving the computational efficiency of the method of moments. If the current solution is not smooth, as is the case near a geometrical singularity, polynomials are not well suited to represent the solution, and an efficiency gain is not realized. Near edge singularities, h-refinement can be better than p-refinement. In the analysis of this chapter, we will therefore focus on smooth scatterers. We will take as a canonical geometry the interior of the flat strip as treated in Section 5.1, neglecting the edge regions of the scatterer and associated current singularities. We will consider only ideal discretizations, so that quadrature error and geometrical discretization error are assumed to be negligible.

8.1 HIGHER-ORDER BASIS FUNCTIONS FOR 2D PROBLEMS

For two-dimensional scattering problems, higher-order polynomials are defined with support on one-dimensional mesh elements. The approximation space for the current solution on the canonical mesh element $[-h/2, h/2]$ consists of $p + 1$ polynomials

$$f_a(x), \quad a = 0, 1, \dots, p, \quad -h/2 \le x \le h/2 \tag{8.1}$$

where a indexes the family of polynomials used to expand the current solution and p is the degree of the highest-order polynomial in the basis set. These polynomials are shifted and scaled if necessary to provide $p+1$ basis functions $f_{n,a}(x)$ on the nth mesh element. The current solution is of the form

$$\hat{J}(x) = \sum_{n=1}^{N} \sum_{a=0}^{p} I_{n,a} f_{n,a}(x) \tag{8.2}$$

where $I_{n,a}$ is an unknown coefficient representing the weight of the ath polynomial basis function and N is the number of mesh elements.

We will define the mesh element density $n_\lambda = \lambda/h$ as before, but for higher-order basis functions, the density of the degrees of freedom (unknowns) increases to $(p + 1)n_\lambda$ because of the use of multiple functions per element to represent the current solution. For the polynomial basis functions f_a, we will consider both orthogonal and interpolatory polynomials.

Legendre polynomials (orthogonal basis). The basis functions on each element consist of the scaled Legendre polynomials

$$f_a(x) = c_a P_a(2x/h) \tag{8.3}$$

where a is the order of the polynomial and $c_a = \sqrt{2a + 1}$ is a normalization constant. The Legendre polynomials $P_a(x)$ are orthogonal with respect to a uniform weighting function on the interval $[-1, 1]$. The order $p = 0$ case is identical to the pulse functions studied previously, and the first-order ($p = 1$) case consists of constant and linear functions on each mesh element.

Lagrange polynomials (interpolatory basis). To define interpolatory polynomials, $p+1$ node points on each mesh element must be chosen. For the element $[-h/2, h/2]$, we will denote the node points as y_a, $a = 0, 1, \ldots, p$, with the restriction $-h/2 \le y_a \le h/2$ so that the nodes are confined to the element. The basis functions are

$$f_a(x) = L_{p,a}(x) \tag{8.4}$$

where $L_{p,a}(x)$ is the ath canonical Lagrange polynomial of order p with respect to the $p + 1$ nodes y_a. $L_{p,a}(x)$ is completely determined by the interpolatory property that the polynomial is unity at the ath node point and zero at the others:

$$L_{p,a}(y_b) = \begin{cases} 1 & a = b \\ 0 & a \ne b \end{cases} \tag{8.5}$$

The canonical polynomials can be given explicitly as

$$L_{p,a}(x) = \prod_{b=0, b \ne a}^{p} \frac{(x - y_b)}{y_a - y_b} \tag{8.6}$$

For these basis functions, the trial solution (8.2) is equal to the coefficient $I_{n,a}$ at $x = x_{n,a}$, where $x_{n,a}$ is the node point y_a mapped to the nth mesh element.

The first-order Lagrange polynomials are closely related to the triangle functions considered in earlier chapters. If we take the x coordinates of the apexes of two adjacent triangle functions to be the mesh element endpoints, and choose the nodes $x_{n,0}$ and $x_{n,1}$ to be the mesh endpoints as well (i.e., $y_0 = -h/2$, $y_1 = h/2$), then the portions of the two triangle functions with support on the element are identical to the order $p = 1$ canonical Lagrange polynomials. The only difference between the triangle functions and the Lagrange polynomial expansion is that with the triangle basis function, the weights of the two linear portions on adjacent mesh elements are constrained to be equal, whereas this continuity across mesh element boundaries is not enforced with the higher-order polynomial basis functions considered here. Since continuity is not enforced, there are two degrees of freedom per mesh element with Lagrange polynomials, and the total number of unknowns is twice as large as with triangle functions. It has been shown that the moment method solution is continuous at mesh element boundaries, even if continuity is not enforced [1]. It will be seen that the solution convergence order is identical for these two related types of first-order basis functions.

8.2 INTERPOLATORY POLYNOMIALS

The goal here is to extend the machinery of Chapter 5 for low-order error analysis to higher-order basis functions. The scatterer is the flat strip of width d with endpoints $(-d/2, 0)$ and $(d/2, 0)$. The mesh consists of N elements having midpoints $x_n = (n - 1/2)h - d/2, n = 1, \ldots, N$, where $N = d/h$. With this regular mesh, the basis functions have the form

$$f_{n,a}(x) = L_{p,a}(x - x_n), \quad x_n - h/2 \le x \le x_n + h/2 \tag{8.7}$$

on the nth mesh element. The node points used to define the canonical Lagrange polynomials on the element are $x_{n,a} = x_n + y_a$.

8.2.1 Discretized Operator Spectrum

The first task is to determine the spectral error introduced by discretization of the TM-EFIE with the higher-order basis functions (8.4). We will consider the case of identical testing and expansion functions (Galerkin's method). The moment matrix associated with a higher-order discretization can be expressed using index notation as

$$Z_{mn,ab} = h^{-1}\langle f_{m,a}, \mathcal{L}f_{n,b}\rangle \tag{8.8}$$

where $\langle \cdot, \cdot \rangle$ is the L^2 inner product. By making use of (5.10), this can be put in the spectral form

$$Z_{mn,ab} = \frac{\eta}{2n_\lambda} \int \frac{d\beta}{\sqrt{1-\beta^2}} e^{-j2\pi\beta(m-n)/n_\lambda} F_a(-\beta) F_b(\beta) \tag{8.9}$$

where $\beta = k_x/k_0$ and $F_a(\beta)$ is the Fourier transform of $f_a(x)$, normalized by h^{-1} as in (5.12).

To accommodate multiple basis functions per element, it is helpful to define an approximate operator that is related to the moment matrix but acts directly on the function space of currents, rather than the discrete space of current samples or unknowns. This approximate operator can be represented as

$$\hat{\mathcal{L}}u = h^{-1} \sum_{m,n=1}^{N} \sum_{a,b=0}^{p} f_{m,a} Z_{mn,ab} \langle f_{n,b}, u \rangle \tag{8.10}$$

This definition involves projecting the function u onto the expansion functions to obtain a finite number of coefficients, applying the moment matrix (8.8), and combining the testing functions using the resulting coefficients, so the operator moves from a function space to a discrete space and back to the function space.

For interpolatory basis functions, there are two ways to perform the projection $\langle f_{n,b}, u \rangle$ from the space of continuous functions to a finite number of basis expansion coefficients. The first is interpolation, as discussed in Section 2.8.1, and the second is L^2 projection, as in Section 2.8.3. For interpolatory basis functions, the analysis is mathematically more natural if the former approach is used. Accordingly, we will take the inner product in (8.8) to be the discrete RMS value where the sample points are the node points $x_{n,a}$. With this choice for the projection used in (8.10), by the interpolatory property of the basis functions we have $\langle f_{n,b}, u \rangle_{\mathrm{RMS}} = u(x_{n,b})$.

We will estimate the eigenvalues of the discretized operator using the normal approximation developed in Section 5.1.1. The operator eigenvalues are approximated by the diagonal elements of (5.1). Replacing the exact operator $\mathcal{L}$ with the approximate operator $\hat{\mathcal{L}}$ and the L^2 inner product with the discrete RMS value leads to the discretized operator eigenvalue estimate

$$\hat{\lambda}_q \simeq N^{-1} \left\langle e^{-j\beta_q k_0 x}, \hat{\mathcal{L}} e^{-j\beta_q k_0 x} \right\rangle_{\mathrm{RMS}(x_{n,a})} \tag{8.11}$$

where $\beta_q = q/D$ is the normalized mode spatial frequency and N here is the number of node points, rather than the number of mesh elements. In Section 5.1, we used L_{qq} and $\hat{L}_{qq}$ to denote eigenvalue estimates obtained using the normal approximation method for the integral operator and moment matrix, respectively, but here for simplicity we will ignore the distinction between the true eigenvalues and the estimates obtained using the normal approximation and denote the operator and moment matrix eigenvalue estimates as λ_q and $\hat{\lambda}_q$.

Evaluating the RMS value in (8.11) and using the interpolatory property of the basis functions leads to

$$\hat{\lambda}_q \simeq N^{-1} \sum_{m,n=1}^{N} \sum_{a,b=0}^{p} e^{j\beta_q k_0 x_{m,a}} Z_{mn,ab} e^{-j\beta_q k_0 x_{n,b}} \tag{8.12}$$

This is a generalization of (5.9) that allows for multiple sample points per mesh element.

We can interpret (8.12) in terms of the matrix transformation given by

$$U_{ma,q} = N^{-1/2} e^{-jk_0 \beta_q x_{m,a}} \tag{8.13}$$

If the nodes points $x_{m,a}$ are evenly spaced, then the transformation is unitary, since

$$[\mathbf{U}^H \mathbf{U}]_{qr} = N^{-1} \sum_{m=1}^{N} \sum_{a=0}^{p} e^{jk_0 \beta_q x_{m,a}} e^{-jk_0 \beta_r x_{m,a}} = \delta_{qr} \tag{8.14}$$

Equation (8.12) can be seen to be the qth diagonal element of the matrix

$$\mathbf{L} = \mathbf{U}^H \mathbf{Z} \mathbf{U} \tag{8.15}$$

As shown in Section 5.1.1, since this unitary transformation leads to a strongly diagonal matrix, the diagonal elements provide eigenvalue estimates.

By following the treatment of Section 5.1, we can obtain the eigenvalue estimate

$$\hat{\lambda}_q \simeq \frac{\eta}{2n_\lambda^2 D} \int_{-\infty}^{\infty} \frac{d\beta}{\sqrt{1-\beta^2}} \frac{\sin^2\left[\pi D(\beta - \beta_q)\right]}{\sin^2\left[\pi(\beta - \beta_q)/n_\lambda\right]} F^{(p)^2}(\beta, \beta_q) \qquad (8.16)$$

where

$$F^{(p)}(\beta, \beta') = \sum_{a=0}^{p} e^{-j\beta' k_0 y_a} F_a(\beta) \qquad (8.17)$$

In deriving (8.16), we have assumed symmetrical node points y_a, so that $F(-\beta, -\beta') = F(\beta, \beta')$.

8.2.2 Interpolation Transfer Function

The function $F^{(p)}(\beta, \beta')$ in (8.17) is the Fourier transform of the interpolated polynomial approximation obtained by sampling the mode $e^{-j\beta' k_0 x}$ at the node points y_a and using the samples in (8.2) as coefficients for the basis functions on one mesh element. We will refer to $F^{(p)}(\beta, \beta')$ as an interpolation transfer function.

Since the transfer function $F^{(p)}(\beta, \beta')$ figures heavily in the following treatment, it will be helpful to understand its properties before moving forward. The Fourier transform of the mode $u(x) = e^{-j\beta' k_0 x}$ on one mesh element scaled by $1/h$ is

$$\frac{1}{h} \int_{-h/2}^{h/2} e^{jk_0 \beta x} u(x)\, dx = \frac{\sin\left[k_0(\beta - \beta')h/2\right]}{k_0(\beta - \beta')h/2} = \text{sinc}[(\beta - \beta')/n_\lambda] \qquad (8.18)$$

Because the polynomial representation is incomplete, for fixed β' the Fourier transform $F^{(p)}(\beta, \beta')$ of the interpolated polynomial approximation to the mode $e^{-j\beta' k_0 x}$ is not exactly equal to the ideal sinc function in (8.18). For $p = 0$, $F^{(0)}(\beta, \beta')$ reduces to $\text{sinc}(\beta/n_\lambda)$, which is identical to (2.58) but quite different from (8.18). As p increases, the interpolated mode becomes closer to the exact mode, and $F^{(p)}(\beta, \beta')$ approaches the exact Fourier transform (8.18). This is illustrated in Figure 8.1. The convergence of $F^{(p)}(\beta, \beta')$ for large p to the exact Fourier transform means that little error is incurred when interpolating the mode with a high-order polynomial representation.

Of particular importance is the amplitude of the original mode $e^{-j\beta' k_0 x}$ in the interpolated polynomial approximation. This is given by

$$F^{(p)}(\beta') \equiv F^{(p)}(\beta', \beta') \qquad (8.19)$$

If $F^{(p)}(\beta')$ is significantly different from unity, then the basis function expansion provides a poor representation of the mode. As p becomes large, we have the limit

$$\lim_{p \to \infty} F^{(p)}(\beta') = 1 \qquad (8.20)$$

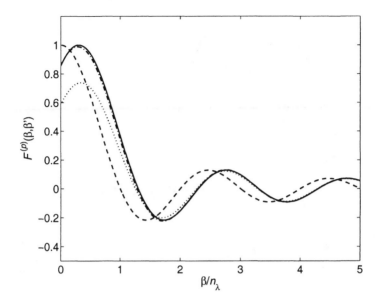

Figure 8.1: Interpolation transfer function $F^{(p)}(\beta, \beta')$ for Lagrange interpolatory polynomial expansion with $\beta'/n_\lambda = 0.3$. Dashed line: $p = 0$. Dotted: $p = 1$. Dash-dot: $p = 2$. Solid line: exact mode Fourier transform, (8.18). As the polynomial order increases, $F^{(p)}(\beta, \beta')$ approaches the exact Fourier transform.

This means that the amplitude of the mode $e^{-j\beta' k_0 x}$ in the interpolated approximation converges to one as the order of the basis set becomes large. Since the interpolatory polynomials add to unity on the element, from (8.17) it follows that we have the same limiting value as h becomes small.

Figure 8.2 shows $F^{(p)}(\beta)$ for several polynomial orders. The improved approximating power of the basis functions with order p is evident as the "passband" of the transfer function becomes flatter as the polynomial order increases and $F^{(p)}(\beta)$ approaches an ideal low-pass filter characteristic. As the order increases and more polynomials are used to expand the current solution on each element, the function $F^{(p)}(\beta)$ becomes closer to unity for small values of β. The mesh Nyquist frequency defined in (2.63) corresponds to $\beta/n_\lambda = 1/2$, from which it can be seen that for orders $p = 2$ and higher, $F^{(p)}(\beta)$ is very close to one over the range of spatial frequencies corresponding to the modeled modes discussed in Section 3.1.2.

To obtain explicit results for the transfer function, we will choose the nodes y_a to be evenly and symmetrically spaced on $[-h/2, h/2]$, with $y_0 = -h/2$ and $y_p = h/2$ if p is odd and $y_{p/2}$ at the element center if p is even. Closed-form expressions for

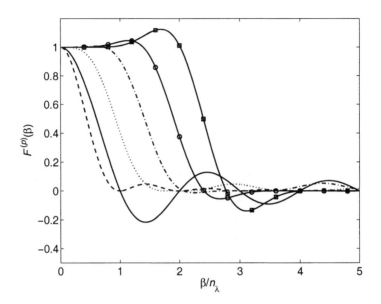

Figure 8.2: $F^{(p)}(\beta)$ for Lagrange interpolatory polynomial expansion. Solid line: $p = 0$. Dashed: $p = 1$. Dotted: $p = 2$. Dash-dot: $p = 3$. Solid/circles: $p = 4$. Solid/squares: $p = 5$.

$F^{(p)}(\beta, \beta')$ are rather lengthy, but the small argument expansions

$$F^{(0)}(\beta) \simeq 1 - \frac{\pi^2 \beta^2}{6n_\lambda^2}$$

$$F^{(1)}(\beta) \simeq 1 - \frac{\pi^2 \beta^2}{3n_\lambda^2}$$

$$F^{(2)}(\beta) \simeq 1 - \frac{\pi^4 \beta^4}{60n_\lambda^4} \tag{8.21}$$

$$F^{(3)}(\beta) \simeq 1 - \frac{\pi^4 \beta^4}{405n_\lambda^4}$$

can be obtained. It is interesting to observe that the exponents do not increase monotonically and increment only for every other order.

From these considerations, we can see that for both p-refinement and h-refinement, the behavior of the transfer function $F^{(p)}(\beta, \beta')$ reflects the improved approximating power of the basis set. We will find that the deviation of $F^{(p)}(\beta, \beta')$ from the ideal Fourier transform of the mode $e^{-j\beta' k_0 x}$ determines the spectral error associated with the discretization.

8.2.3 Spectral Error

The integrand of (8.16) has local maxima at $\beta_{q,s} = \beta_q + sn_\lambda, s = 0, \pm 1, \pm 2, \ldots$. Expanding the integrand about each local maximum and retaining the leading-order term yields the spectral estimate

$$\hat{\lambda}_q \simeq \sum_{s=-\infty}^{\infty} \lambda_{q+sN} F^{(p)^2}(\beta_{q,s}, \beta_q) \tag{8.22}$$

where λ_q represents the operator eigenvalue estimate (5.5). The relative spectral error is

$$E_q^{(p)} = \frac{\hat{\lambda}_q - \lambda_q}{\lambda_q} = F^{(p)^2}(\beta_q) - 1 + \frac{1}{\lambda_q} \sum_{s\neq 0} \lambda_{q+sN} F^{(p)^2}(\beta_{q,s}, \beta_q) \tag{8.23}$$

As before, we can decompose the spectral error into two contributions,

$$E_q^{(p)} = E_q^{(p,1)} + E_q^{(p,2)} \tag{8.24}$$

where the projection error $E_q^{(p,1)}$ is defined to be $F^{(p)^2}(\beta_q) - 1$ and the aliasing error is the remaining sum in (8.23).

8.2.4 Current Solution Error

Armed with spectral error estimates, we can now analyze the current solution error associated with smooth scatterer regions away from geometrical singularities. For the flat strip, this is the interior region $\tilde{C}$ with small regions near the edges omitted from the error computation as in Sections 5.1 and 5.2. For a plane incident wave, using the normal EFIE operator approximation the current solution can be approximated as in Chapter 5 by

$$J(x) \simeq \lambda_q^{-1} e^{jk_0 x \cos \phi} \tag{8.25}$$

where $q = D \cos \phi$ and the fractional part of $D \cos \phi$ is unimportant for large D.

To find the numerical solution obtained using the method of moments, we can transform the linear system (2.27) to

$$\mathbf{U}^H \mathbf{Z} \mathbf{U} \mathbf{x} = \mathbf{U}^H \mathbf{v}^i \tag{8.26}$$

with the unitary matrix given by (8.13). Using (8.17) and (8.19), it can be shown that the right-hand side has one nonzero element given by

$$[\mathbf{U}^H \mathbf{v}^i]_q = F^{(p)}(\beta_q) \tag{8.27}$$

where $\beta_q = \cos \phi$. The approximate numerical solution can then be estimated as

$$\hat{J}(x) \simeq \hat{\lambda}_q^{-1} F^{(p)}(\beta_q) e^{jk_0 x \cos \phi} \tag{8.28}$$

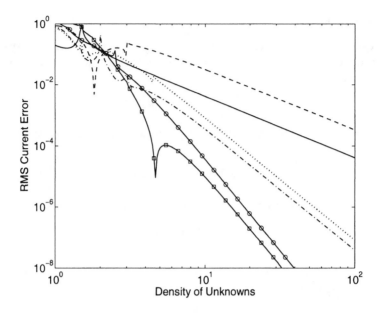

Figure 8.3: Relative current solution error estimate (8.30) for Lagrange polynomial basis functions. Error is measured in the discrete RMS norm at the node points used to define the Lagrange polynomials. Angle of incidence: $\phi^{inc} = 60°$. Integral equation: EFIE. Polarization: TM. Moment matrix element integration: exact. Scatterer: interior of flat strip. The independent variable is the total density of the degrees of freedom per wavelength, $(p+1)n_\lambda$. Solid line: $p = 0$. Dashed: $p = 1$. Dotted: $p = 2$. Dash-dot: $p = 3$. Solid/circles: $p = 4$. Solid/squares: $p = 5$.

In deriving this expression, we have ignored an additional interpolation error term introduced when the current solution (8.2) is expanded using the basis functions together with the coefficients from the solution of the linear system (8.26). Since the interpolatory basis functions are equal to one at the node points $x_{n,a}$, we can use the discrete RMS norm to measure the current solution error and avoid this additional source of error. Since the RMS error was used in earlier chapters for low-order basis functions, this means that the higher-order results in this section can be compared directly to the lower-order cases considered previously.

Using these approximations for the exact and numerical current solutions, we can estimate the current solution error on the interior of the flat strip. The discrete RMS relative interior current solution error is

$$\text{Err}_{\text{RMS}(\tilde{C})} = \frac{\|J - \hat{J}\|_{\text{RMS}(\tilde{C})}}{\|J\|_{\text{RMS}(\tilde{C})}} \simeq \left| \frac{\hat{\lambda}_q - \lambda_q F^{(p)}(\beta_q)}{\hat{\lambda}_q} \right| \tag{8.29}$$

which has the same form as (5.29). Assuming that the spectral error is small relative

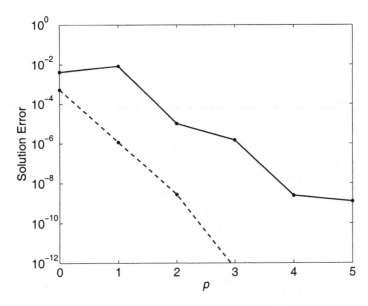

Figure 8.4: Relative current solution error estimate (8.30) (solid line) and scattering amplitude error estimate (8.31) (dashed line) for Lagrange polynomial basis functions as a function of order. Angle of incidence: $\phi^{inc} = 60°$. Integral equation: EFIE. Polarization: TM. Moment matrix element integration: exact. Scatterer: interior of flat strip. Mesh element density: $n_\lambda = 10$.

to λ_q (which holds for nonresonant scatterers), the relative error can be approximated as

$$\text{Err}_{\text{RMS}(\tilde{C})} \simeq \left| E_q^{(p,2)} + F^{(p)}(\beta_q) \left[F^{(p)}(\beta_q) - 1 \right] \right| \tag{8.30}$$

As with lower-order basis functions, projection error caused by the testing functions cancels in the current solution. In the projection error term of the spectral error (8.23), the transfer function $F^{(p)}(\beta_q)$ is squared, but since $F^{(p)}(\beta_q) \left[F^{(p)}(\beta_q) - 1 \right] \simeq F^{(p)}(\beta_q) - 1$, the exponent effectively reduces by one and only the projection error due to the expansion functions remains in the current solution error.

This error estimate is shown in Figure 8.3 as a function of the density of degrees of freedom per wavelength. The asymptotic order of the current solution error in the mesh element length $h = \lambda/n_\lambda$ can be obtained from the projection error term $F^{(p)}(\beta_q) - 1$ of (8.30). From (8.21), the discrete RMS current error order is $p + 2$ for p even, and $p + 1$ for p odd. For the $p = 1$ Lagrange polynomials, the order is h^2, which is identical to the current solution convergence rate obtained in earlier chapters for triangle basis functions.

Figure 8.4 shows the RMS current error (8.30) as a function of polynomial order. The exponential convergence of the interior current solution error with respect to polynomial order is evident in the figure, since the order is shown on a linear scale and the logarithmic error falls off roughly linearly. The increase in current error from $p = 0$ to $p = 1$ matches the results of Chapters 3 and 5 that projection error for triangle functions is larger than that of pulse functions. This is evident, for example, in (5.20).

8.2.5 Scattering Amplitude Error

To compute the numerical scattering amplitude (2.45) with higher-order basis functions, a plane wave in the scattering direction must be projected onto the basis functions using a generalization of (2.46) allowing for multiple polynomials on each mesh element. This projection leads to an additional factor of $F^{(p)}(\beta)$ in the scattered field that cancels a like factor in the projection error term of (8.23). As a result, the scattering amplitude error is determined only by the aliasing error term of (8.23). This is another manifestation of the variational property discussed in Section 3.2. The resulting relative specular scattering amplitude error estimate is

$$\frac{|S(\phi) - \hat{S}(\phi)|}{|S(\phi)|} \simeq \left| E_q^{(p,2)} \right| = \left| \frac{1}{\lambda_q} \sum_{s \neq 0} \lambda_{q+sN} F^{(p)2}(\beta_{q,s}, \beta_q) \right| \tag{8.31}$$

This result neglects the error caused by current singularities at the edges of the strip, so it must be viewed as a scattering amplitude error estimate for smooth scatterers. The error estimate is shown in Figure 8.5, from which it is evident that the scattering amplitude error is of order $2p + 3$.

The terms of the summation over s in (8.31) fall off as $1/s$ or faster, and the order of the sum with respect to n_λ^{-1} is the same as that of the $s = 1$ term of the sum. With this simplification, the error becomes

$$\left| E_q^{(p,2)} \right| \simeq \frac{\eta}{2|\lambda_q|n_\lambda} F^{(p)2}(\beta_q + n_\lambda, \beta_q) \tag{8.32}$$

where we have used (5.5). For the irregular cases in Section 4.5.1.2, cancellations in the summation cause this reasoning to break down, but for higher-order discretizations, the approximation is valid.

To understand these results, we recall that $F^{(p)}(\beta, \beta_q)$ converges to the Fourier transform of the mode $e^{-j\beta_q k_0 x}$ as p increases. The exact Fourier transform (8.18) vanishes identically at $\beta = \beta_q + n_\lambda$, so (8.32) approaches zero for large p. Moreover, the exact Fourier transform is zero at $\beta_{q,s} = \beta_q + sn_\lambda$, $s \neq 0$, so all terms in the summation in (8.31) are negligible as well. For small p, $F^{(p)}(\beta_q + sn_\lambda, \beta_q)$ is nonzero due to error caused by interpolation of the mode $e^{-j\beta_q k_0 x}$ by the Lagrange polynomials basis functions. Thus, the scattering amplitude error as well as the current error are determined by the deviation of the interpolation transfer function from the exact mode Fourier

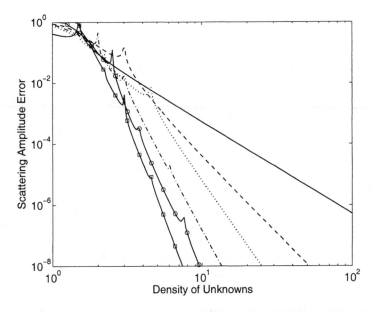

Figure 8.5: Relative scattering amplitude error estimate (8.31) for Lagrange polynomial basis functions. Angle of incidence: $\phi^{\text{inc}} = 60°$. Integral equation: EFIE. Polarization: TM. Moment matrix element integration: exact. Scatterer: interior of flat strip. Solid line: $p = 0$. Dashed: $p = 1$. Dotted: $p = 2$. Dashdot: $p = 3$. Solid/circles: $p = 4$. Solid/squares: $p = 5$.

transform. The current error is determined by the deviation of the amplitude of the mode in the interpolated polynomial approximation from unity, and the scattering amplitude error is determined by the amplitudes of the aliases of the mode at spatial frequencies $k_0 \beta_q + s k_0 n_\lambda$, $s \neq 0$.

The convergence order of the estimate (8.31) can be obtained in a simple way using the smoothness arguments developed for the regular discretizations considered in Section 4.5.1.1. The Fourier transform of a continuous polynomial function of order p is asymptotically $O(k^{-p-1})$. The TM-EFIE kernel has a 1D Fourier transform of order $O(k^{-1})$. Combining these factors for the kernel and testing and expansion functions leads to a combined asymptotic behavior of $O(k^{-2p-3})$, which corresponds to an aliasing error of order h^{2p+3} or n_λ^{-2p-3}. For the TE-EFIE, derivative operators decrease the smoothness of the kernel so that the Fourier transform of the kernel is $O(k)$, and the aliasing error order decreases to h^{2p+1}. For the MFIE, a smoothness order argument based on the identity term in the operator leads to an aliasing error of order h^{2p+2}.

These predicted convergence rates agree with the error analyses of low-order basis functions of earlier chapters. For the $p = 1$ polynomial basis functions, the scattering amplitude solution convergence orders predicted above for Galerkin testing are h^3,

h^4, and h^5 for the TE-EFIE, MFIE, and TM-EFIE, respectively. These estimates match those given in the $p + p' = 2$ column of Table 4.1 for triangle testing and expansion functions.

The analysis of this section was carried out for Galerkin testing. For point matching, the Fourier transform of the delta testing function is $O(1)$ with respect to k, and the convergence rates decrease to h^p, h^{p+1}, and h^{p+2} for the TE-EFIE, MFIE, and TM-EFIE, respectively. It should be observed that these estimates do not always hold for $p < 3$, since some of the lowest-order discretizations are irregular as described in Section 4.5.1.2.

8.3 ORTHOGONAL POLYNOMIALS

For orthogonal basis functions beyond order $p = 0$, the interpolatory property used in earlier chapters and in Section 8.2 is not available, and in order to develop solution error estimates, the treatment must be modified by replacing the RMS error measure with the L^2 error for the current solution.

8.3.1 Discretized Operator Spectrum

For orthogonal basis functions, we will define the approximate operator (8.10) with the L^2 inner product. With this interpretation, the approximate operator is now equivalent to orthogonal projection of a continuous function onto the polynomial expansion functions, leading to a finite number of coefficients to which the moment matrix is applied, after which a continuous function is finally obtained from a linear combination of the testing functions. The basis functions appear *four* times in the approximate operator—twice in the moment matrix, and twice in moving from the continuous function space to a discrete space and back again.

To determine the spectrum of the approximate operator, we can use (8.11) with the RMS value replaced by the L^2 inner product. This leads to the eigenvalue estimate

$$\hat{\lambda}_q \simeq N^{-1} \left\langle e^{-j\beta_q k_0 x}, \hat{\mathcal{L}} e^{-j\beta_q k_0 x} \right\rangle_{L^2} \tag{8.33}$$

Inserting the definition of $\hat{\mathcal{L}}$ and following the derivation of the preceding section leads to an eigenvalue estimate of the same form as (8.16), but with $F^{(p)}(\beta, \beta')$ redefined according to

$$F^{(p)}(\beta, \beta') = \sum_{a=0}^{p} F_a(\beta) F_a(-\beta') \tag{8.34}$$

This is a Fourier representation of the projection operator from the space of L^2 functions onto the trial subspace spanned by the basis polynomials.

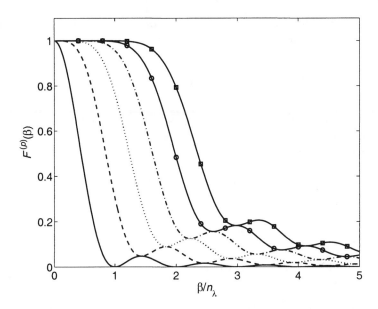

Figure 8.6: $F^{(p)}(\beta)$ for the Legendre polynomial expansion. Solid line: $p = 0$. Dashed: $p = 1$. Dotted: $p = 2$. Dash-dot: $p = 3$. Solid/circles: $p = 4$. Solid/squares: $p = 5$.

8.3.2 Projection Transfer Function

To understand $F^{(p)}(\beta, \beta')$ as defined by (8.34), we observe that this is the Fourier transform of the mode $e^{-j\beta' k_0 x}$ after projection onto the basis function subspace. More specifically, the mode is first projected onto the basis functions using the L^2 inner product to obtain a finite number of coefficients, after which the coefficients are used to form a linear combination of the same basis functions. Whereas (8.17) is associated with polynomial interpolation of a continuous function, (8.34) is determined by L^2 projection onto a polynomial space. $F^{(p)}(\beta, \beta')$ can also be viewed as the Fourier representation of the approximate completeness relation associated with the orthogonal polynomial basis. We will refer to (8.34) as a projection transfer function.

As with the interpolation transfer function defined previously, (8.34) converges to the Fourier transform of the mode $e^{-j\beta' k_0 x}$ as the polynomial order of the basis functions increases. For small values of the order p, $F^{(p)}(\beta, \beta')$ is significantly different from the exact Fourier transform. In particular, $F^{(p)}(\beta') = F^{(p)}(\beta', \beta')$ defined by (8.19) is different from unity, so that projection introduces a scale factor in the amplitude of the mode $e^{-j\beta' k_0 x}$ that is analogous to the scale factor in (2.65). For p-refinement and h-refinement, $F^{(p)}(\beta')$ has the same limiting behavior as in (8.20).

For the Legendre polynomial basis functions, we can give an explicit formula for the projection transfer function. Since the Fourier transform of a Legendre polynomial is proportional to a spherical Bessel function, (8.34) evaluates to

$$F^{(p)}(\beta, \beta') = \sum_{a=0}^{p} (2a + 1) j_a(\pi\beta/n_\lambda) j_a(\pi\beta'/n_\lambda) \tag{8.35}$$

where $j_a(x)$ is the spherical Bessel function of order a. The improvement in the approximation power as p increases can be seen analytically by expanding $F^{(p)}(\beta)$ for small β, which leads to

$$F^{(0)}(\beta) \simeq 1 - \frac{\pi^2\beta^2}{3n_\lambda^2}$$

$$F^{(1)}(\beta) \simeq 1 - \frac{\pi^4\beta^4}{45n_\lambda^4} \tag{8.36}$$

$$F^{(2)}(\beta) \simeq 1 - \frac{\pi^6\beta^6}{1575n_\lambda^6}$$

The improved solution convergence obtained with p-refinement is evident in the decrease of the constant and the increasing order of these expansions with p.

The behavior of $F^{(p)}(\beta, \beta')$ for fixed β' is similar to Figure 8.1 for Lagrange polynomials. $F^{(p)}(\beta)$ is shown in Figure 8.6 for several values of the order p. As the order of the basis set increases and more polynomials are used to expand the current solution on each element, the function $F^{(p)}(\beta)$ becomes closer to unity for small values of β. For orders $p = 2$ and higher and modeled modes with spatial frequency below the mesh Nyquist frequency $k_0 n_\lambda/2$, or $\beta/n_\lambda < 1/2$, $F^{(p)}(\beta)$ is close to one, indicating small projection error. Because the projection transfer function converges to something analogous to a rectangular low-pass spatial filter characteristic, higher-order basis functions reduce the spectral error associated with discretization of the integral operator.

8.3.3 Spectral Error

The approximate operator eigenvalues for orthogonal basis functions are given by (8.22), but with $F^{(p)}(\beta, \beta')$ defined as in (8.34), and the spectral error is identical to (8.23) with the same reinterpretation of the projection transfer function.

8.3.4 Current Solution Error

In analyzing the current solution error for higher-order orthogonal basis functions, we must make a departure from the approach used in earlier chapters. The spectral error (8.23) with the projection transfer function (8.34) does contribute to the current

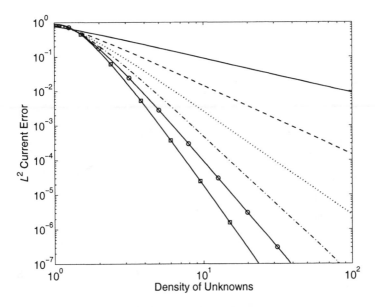

Figure 8.7: Relative L^2 norm current solution error estimate (8.39) for Legendre polynomial basis functions. Angle of incidence: $\phi^{inc} = 60°$. Integral equation: EFIE. Polarization: TM. Moment matrix element integration: exact. Scatterer: interior of flat strip. The independent variable is the density of degrees of freedom per wavelength $(p + 1)n_\lambda$. Solid line: $p = 0$. Dashed: $p = 1$. Dotted: $p = 2$. Dash-dot: $p = 3$. Solid/circles: $p = 4$. Solid/squares: $p = 5$.

solution error, but with a noninterpolatory basis, we cannot measure the discrete RMS current error at special sample points where the interpolation error is zero. Here, we must measure the current error using the L^2 norm.

The current solution error for the interior of the flat strip without edge singularity effects can be estimated simply by projecting the physical optics current mode onto the basis functions and computing the L^2 error, as was done in deriving (2.66) for a single basis function per mesh element. Projecting the mode $J(x) = e^{-j\beta k_0 x}$ onto the basis functions leads to the coefficients

$$I_{n,a} = \int e^{-j\beta k_0 x} f_{n,a}(x)\, dx \tag{8.37}$$

The projected mode is then expanded using the basis functions according to (8.2) to obtain the continuous function

$$\hat{J}(x) = \sum_{n,a} I_{n,a} f_{n,a}(x) \tag{8.38}$$

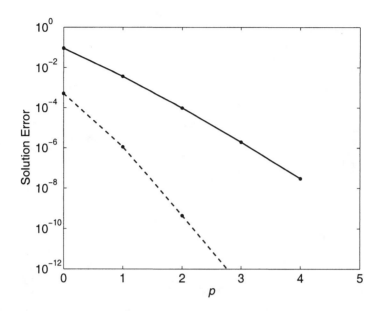

Figure 8.8: Relative current solution error estimate (8.39) (solid line) and scattering amplitude error estimate (8.31) (dashed line) for Legendre polynomial basis functions as a function of order. Angle of incidence: $\phi^{\text{inc}} = 60°$ ($\beta = 0.5$). Integral equation: EFIE. Polarization: TM. Moment matrix element integration: exact. Scatterer: interior of flat strip. Mesh element density: $n_\lambda = 10$.

It can be shown that the relative L^2 error is

$$\frac{\|\hat{J} - J\|}{\|J\|} = \left|1 - F^{(p)}(\beta)\right|^{1/2} \tag{8.39}$$

which generalizes (2.66) to higher-order basis functions.

This error estimate is shown in Figure 8.7, where it can be seen that the L^2 error is of order h^{p+1}. We have ignored the additional perturbation to the current solution caused by the spectral error, but for the Legendre polynomial basis functions, the spectral error has order h^{2p+2} and is negligible in comparison to the dominant error contribution in (8.39). A similar observation for the $p = 0$ case was made in Section 2.8, where it was found that L^2 error was dominated by interpolation error and had order h for pulse functions ($p = 0$). Figure 8.8 shows the error estimate (8.39) as a function of order.

8.3.5 Scattering Amplitude Error

For orthogonal basis functions, we have the same cancellation of projection error due to variationality that occurs with interpolatory polynomials, from which it follows that the scattering amplitude error is determined by the aliasing component of the spectral error. This leads to a relative specular scattering amplitude error estimate identical to (8.31), but with $F^{(p)}(\beta, \beta')$ given by (8.34). As before, the estimate neglects the error caused by current singularities at the edges of the strip. The scattering amplitude error is similar to that observed in Figure 8.5 with respect to the number of degrees of freedom. The estimate is shown in Figure 8.8 as a function of polynomial order. The convergence rate of the scattering amplitude solution is h^{2p+3}, which is identical to that observed with Lagrange polynomials. The $p = 0$ Legendre polynomials basis with Galerkin testing is identical to the case of pulse expansion and testing functions, for which the scattering amplitude solution convergence rate is h^3 as given by the $p + p' = 0$ entry for the TM-EFIE in Table 4.1.

8.4 3D PROBLEMS

For 3D scattering problems, the flexibility in choosing higher-order basis functions is greater than for 2D problems, because the polynomial basis functions on surface mesh elements are two-dimensional and the polynomial orders for each dimension can be chosen separately. Because of the continuity properties of physical fields and currents, however, the number of degrees of freedom with a full higher-order polynomial vector expansion is greater than is needed. Typically, a subset of the full space of vector fields with polynomial coefficients is chosen.

The basis function smoothness at mesh element edges can be used to classify these subsets of vector basis functions [2]. There are two main classes of vector basis functions, the divergence conforming basis functions with finite divergence and possibly discontinuous tangential components between adjacent basis functions, and curl conforming basis functions that impose tangential continuity and may have discontinuous normal components. Divergence conforming bases include:

> *Constant normal, linear tangential (CN/LT):* These basis functions have a constant normal component and linear tangential component on mesh element edges. For triangular elements, these are the Rao-Wilton-Glisson (RWG) basis functions [3], and are the most commonly used basis functions for electromagnetic surface integral equations. On a rectangular mesh, the CN/LT vector functions are the rooftop basis functions analyzed above. CN/LT functions combine the one-dimensional pulse and triangle functions.

> *Linear normal, linear tangential (LN/LT):* At element edges, both the normal and tangential components vary linearly. We use LN/LT and LT/LN, respectively, to distinguish the divergence conforming and curl conforming vector basis functions with linear normal and linear tangential components.

> *Linear normal, quadratic tangential (LN/QT):* At element edges, the normal

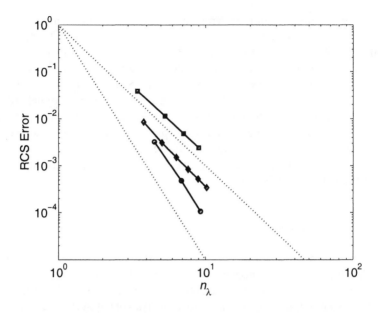

Figure 8.9: RCS error for a PEC sphere with radius 0.5 m at 300 MHz. Integral equation: EFIE. Discretization: ideal, with high-accuracy numerical integration and curved mesh elements. Diamonds: CN/LT (RWG). Squares: LN/LT. Circles: LN/QT. In all cases, the vector basis functions are divergence conforming, and the testing and expansion functions are identical (Galerkin's method). The dotted lines have slopes of −3 and −5.

component of each basis function is linear and the tangential component is quadratic.

Common curl conforming bases include:

Constant tangential, linear normal (CT/LN): These are Nedelec type edge elements [4].

Linear tangential, linear normal (LT/LN).

Linear tangential, quadratic normal (LT/QN).

These basis sets are listed in order of increasing overall polynomial order.

For these basis functions, the convergence order can be predicted using the analysis of Sections 4.5.1.1 and 7.2. For regular discretizations, the contribution of the first eigenvalue alias to the spectral error can be used as a scattering amplitude or RCS error estimate. The first alias can be estimated from the products of the Fourier transforms of the testing functions, expansion functions, and the operator kernel evaluated at the mesh Nyquist frequency (2.63). Since the mesh Nyquist frequency is proportional to

h^{-1}, a Fourier transform which decays asymptotically as $k^{-\alpha}$, $k \to \infty$, corresponds to a scattering amplitude or RCS error of order h^{α}.

We will consider the LN/QT basis with the EFIE as an example. Assuming Galerkin testing and an ideal implementation of the method of moments, the convergence order argument of Section 7.2 becomes

$$Z_{mn} = \frac{jk_0\eta}{h^2} \iint d\mathbf{r} \, d\mathbf{r}' \underbrace{g(\mathbf{r},\mathbf{r}')}_{O(k^{-1})} \left[\underbrace{\mathbf{f}_m(\mathbf{r})}_{O(k^{-2})} \cdot \underbrace{\mathbf{f}_n(\mathbf{r}')}_{O(k^{-2})} + k_0^{-2} \underbrace{\nabla \cdot \mathbf{f}_m(\mathbf{r})}_{O(k^{-2})} \underbrace{\nabla' \cdot \mathbf{f}_n(\mathbf{r}')}_{O(k^{-2})} \right] \quad (8.40)$$

The Fourier transform of the vector basis function is dominated by the linear part of the basis, which leads to a decay rate of k^{-2}. The divergence of a quadratic function is linear, which also corresponds to an $O(k^{-2})$ Fourier transform in the hypersingular term of the EFIE. The overall order of the product of the Fourier transforms of the expansion and testing functions and the kernel is $O(k^{-5})$, which corresponds to an RCS convergence rate of h^5.

8.4.1 Numerical Results

To verify the solution error behavior for the method of moments with higher-order vector basis functions, we will present numerical results for the RCS error. Figure 8.9 shows the RCS error for the EFIE with an ideal discretization (curved mesh elements and accurate integration of moment matrix elements). All the vector basis functions considered are divergence conforming. The scatterer is a PEC sphere with radius $a = 0.5\,\lambda$. The order of the error is the same for the LN/LT and CN/LT bases, although error with the CN/LT basis is smaller. For the LN/QT basis, the order of the polynomial basis is sufficiently high for both the normal and tangential components that the RCS convergence rate improves to fifth order. This is the same as the convergence rate for the $p + p' = 2$ entry for the TM-EFIE in Table 4.1.

As considered at length in Section 4.5, the MFIE can be less accurate than the EFIE, despite the smoothness of the integral part of the operator and the simplicity of discretizing the identity term. This is reflected in the RCS error results for the MFIE with Galerkin testing shown in Figure 8.10, since the fifth-order error observed for the EFIE is not achieved with the MFIE. The fourth-order convergence rate for the LT/LN basis functions matches the fourth-order rates of the $p + p' = 2$ entry in Table 4.1 for the 2D MFIE. The linear dependence of the basis is similar to the triangle function, and triangle basis functions with Galerkin testing leads to the h^4 convergence rate given in Table 4.1 for order $p + p' = 2$. For that case, the order argument given in Section 4.5.1.1 indicates that the scattering amplitude error is dominated by aliasing error caused by the identity part of the MFIE operator.

RCS error is shown in Figure 8.11 for the MFIE with point matching instead of

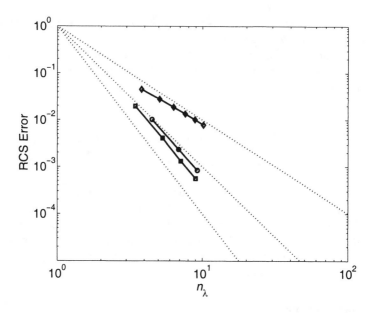

Figure 8.10: RCS error for a PEC sphere with Galerkin's method. Integral equation: MFIE. Discretization: ideal, with high-accuracy numerical integration and curved mesh elements. Squares: LT/LN. Diamonds: CT/LN. Circles: LT/QN. In all cases, the vector basis functions are curl conforming, and the testing and expansion functions are identical (Galerkin's method). The dotted lines have slopes of −2, −3, and −4.

Galerkin testing. The LT/LN basis with point matching is a regular case, since the second-order scattered field convergence rate can be obtained from a smoothness order argument, with point testing contributing an $O(1)$ factor, the identity term $O(1)$, and an $O(h^2)$ factor for the linear expansion functions. The error for the CT/LN and LT/QN bases is not significantly affected by the use of point matching. For CT/LN or LT/QN expansion functions with point matching, the simple operator and basis function smoothness argument predicts a convergence rate of h (first order) for CT/LN and h^2 for LT/QN, but the observed convergence rates are h^2 and h^3, respectively. These must be considered irregular cases, since the smoothness order argument predicts a poorer convergence rate than is observed. Since the computational cost of the method of moments is reduced by the elimination of the testing integral, point matching for the CT/LN and LT/QN bases is more efficient for the same solution accuracy.

One interesting observation is that RCS accuracy is better for the EFIE with the CN/LT basis than with the LN/LT basis (see Figure 8.9), even though the CN/LT basis is a subset of the LN/LT basis. Similarly, the MFIE is slightly more accurate with the LT/LN basis than with the LT/QN basis. The observed convergence rates for the pairs of basis functions are identical, but surprisingly the absolute error is smaller for

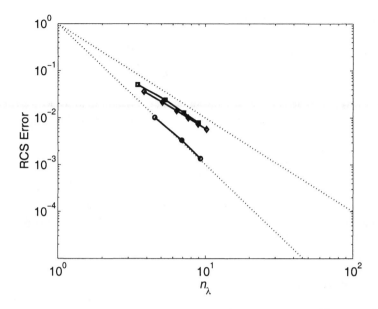

Figure 8.11: RCS error for a PEC sphere with point matching. Integral equation: MFIE. Discretization: ideal, with high-accuracy numerical integration and curved mesh elements. Squares: LT/LN. Diamonds: CT/LN. Circles: LT/QN. In all cases, the vector basis functions are curl conforming. The dotted lines have slopes of −2 and −3.

the lower-order basis. Explanation of this phenomenon must await a more detailed analysis of MoM accuracy with higher-order vector basis functions.

The RCS error order argument illustrated in (8.40) also agrees with empirical convergence rates for higher-order basis functions observed by others. Fink [5] computed RCS errors for a PEC sphere with conformal mesh elements and Galerkin testing. The basis functions were obtained by multiplying the RWG functions with modified Silvester polynomials of order p, so that the $p = 0$ case corresponded to the RWG basis and higher-order bases were complete to order p. If we consider the RWG function to be a combination of a pulse function and a triangle function, the basis functions obtained by multiplying an RWG function by order p Silvester polynomials have an effective polynomial order of p and contribute a factor of h^{p+1} to the RCS error through the first term inside the square brackets of (8.40). The divergence operators in the hypersingular term act on the "triangle" or linear part of the basis functions, so that the basis is effectively one order smoother before the derivative, and the divergence $\nabla \cdot \mathbf{f}$ also contributes a factor of h^{p+1} to the RCS error. The Green's function contributes a factor of h. Combining the factors leads to the RCS error order estimate

$$\text{RCS Error (EFIE)} \sim \underbrace{h^{p+1}}_{\text{f,Testing}} \underbrace{h^1}_{g} \underbrace{h^{p+1}}_{\text{f,Expansion}} + \underbrace{h^{p+1}}_{\nabla \cdot f} \underbrace{h^1}_{g} \underbrace{h^{p+1}}_{\nabla \cdot f} = h^{2p+3} \tag{8.41}$$

For the MFIE, as argued previously, the dominant contribution to the RCS error arises from the identity term of the operator, so the error order analysis becomes

$$\text{RCS Error (MFIE)} \sim \underbrace{h^{p+1}}_{\text{f,Testing}} \underbrace{h^0}_{\mathcal{I}} \underbrace{h^{p+1}}_{\text{f,Expansion}} = h^{2p+2} \tag{8.42}$$

The estimates (8.41) and (8.42) match the numerical results for the EFIE and MFIE with high-accuracy moment matrix integrations and curved mesh elements reported in [5, Figures 3-20 and 3-23].

8.5 Summary

We have considered the problem of obtaining solution error estimates for the method of moments with higher-order polynomial basis functions. By allowing for multiple expansion and testing functions on each mesh element, the spectral error concept developed in earlier chapters can be extended to accommodate higher-order basis functions. The most significant new aspect of the solution error analysis for higher-order bases is the introduction of a transfer function associated with interpolation for Lagrange polynomials or projection of a continuous function onto the basis subspace for orthogonal polynomials. As the order of the polynomial basis increases, this transfer function approaches an ideal low-pass filter characteristic, so that a mode with any spatial frequency between zero and the mesh Nyquist frequency can be accurately represented in the basis subspace. This transfer function determines both the current and scattering amplitude solution errors.

The error estimates in this chapter were for the interior of the flat strip, ignoring error caused by edge singularities. This implies that the solution error convergence orders are valid for smooth scatterers, along the lines of the discussion in Section 3.5. For the TM-EFIE and Legendre polynomial basis functions, the L^2 current solution error for smooth scatterers is of order h^{p+1} and the scattering amplitude error is of order h^{2p+3}. For Lagrange polynomials, the RMS current error is of order h^{p+1} for p odd and h^{p+2} for p even, and the scattering amplitude error is of order h^{2p+3}. The scattering amplitude error has the same order for both types of basis functions. The order $p = 1$ Lagrange polynomial basis with Galerkin testing corresponds to the $p + p' = 2$ entry in Table 4.1, which for the TM-EFIE has order h^5, as expected. Using smoothness order arguments, the convergence rates h^{2p+2} and h^{2p+1} with orthogonal polynomials for the MFIE and TE-EFIE, respectively, are predicted for scattering amplitudes and RCS.

All of these results assume an ideal implementation of the method of moments, with conformal mesh elements or a flat scatterer and exact integration of the moment matrix elements. If an insufficiently accurate integration rule for moment matrix elements is used, solution errors will be worse than predicted by the ideal analysis given here.

For 3D problems, the situation is more complex, because there are additional degrees of freedom in choosing higher-order vector basis functions. In most cases, however, smoothness order arguments predict the observed RCS error convergence rates. With Galerkin testing the RCS convergence rate is h^{2p+3} for the EFIE and h^{2p+2} for the MFIE, as long as the part of the basis function on which the divergence operators in the EFIE act are at least one order smoother than p.

There is a close parallel between the 2D and 3D error behaviors. For the 2D and 3D versions of the method of moments, the MFIE error estimates are identical. For the EFIE, the 2D and 3D error convergence rates also have a perfectly intuitive relationship. An order p vector basis function constructed so that the divergence operator acts on a smoother shape function of order $p + 1$ is equivalent to choosing a scalar basis function for the TE-EFIE that is one order smoother than the basis function for the TM-EFIE. The additional smoothness counteracts the action of the derivatives in the hypersingular term of the EFIE and the convergence orders for both 2D polarizations are h^{2p+3}, which is the same as the observed 3D convergence order.

References

[1] A. F. Peterson and M. M. Bibby, "High-order numerical solutions of the MFIE for the linear dipole," *IEEE Trans. Ant. Propag.*, vol. 52, pp. 2684–2691, Oct. 2004.

[2] A. F. Peterson, R. Mittra, and S. L. Ray, *Computational Methods for Electromagnetics*. New York: IEEE Press, 1998.

[3] S. M. Rao, D. R. Wilton, and A. W. Glisson, "Electromagnetic scattering by surfaces of arbitrary shape," *IEEE Trans. Ant. Propag.*, vol. 30, pp. 409–418, May 1982.

[4] J. C. Nedelec, "Mixed finite elements in R3," *Num. Math.*, vol. 35, pp. 315–341, 1980.

[5] P. W. Fink, *Higher Order Modeling in the BEM and Hybrid BEM/FEM Formulations*. PhD thesis, University of Houston, 2002.

Chapter 9

Operator Spectra and Iterative Solution Methods

Moment method analysis of electrically large radiation and scattering problems can require so many degrees of freedom that direct linear system solution methods are impractical, due to the high computational cost of matrix factorization. To reduce the time required for the linear system solution, iterative algorithms can be employed. For the largest problems, the number of unknowns can be so high that filling the moment matrix is not feasible. Fast methods such as the multilevel fast multipole algorithm (MLFMA) [1] can be used to compute matrix-vector multiplications (matvecs) indirectly, without actually filling the moment matrix. The matvec operation must be used in conjunction with an iterative algorithm to obtain a current solution. For these reasons, iterative linear system solution algorithms are widely used in computational electromagnetics.

There are two main classes of iterative algorithms. The first consists of stationary iterations based on matrix splittings, such as the Jacobi iteration. These algorithms have been used sporadically in conjunction with electromagnetic integral equation solvers [2, 3]. The second class is the nonstationary iterations, the most important of which are methods of the conjugate gradient (CG) family, or Krylov subspace iterations [4]. Both stationary and nonstationary iterations are used in computational electromagnetics [5], but the Krylov subspace family is more common. Stationary iterations perform well for some types of linear systems, but in general are less robust than the Krylov subspace iterations. For these reasons, we will restrict attention here to methods of the later class.

9.1 KRYLOV SUBSPACE ALGORITHMS

The most basic Krylov subspace linear system solution algorithm is the conjugate gradient (CG) algorithm, which can be used for symmetric positive definite (SPD) matrices. Since the linear systems that arise from the method of moments for electromagnetic surface integral equations are non-SPD, the conjugate gradient method can be applied to the normal forms of the linear system (CGNE or CGNR). Other members of the family of Krylov subspace iterations include:

- Biconjugate gradient (BCG);

- Biconjugate gradient-stabilized (BCG-stab);

- Generalized minimum residual (GMRES);

- Quasi-minimum residual (QMR);

- Transpose-free quasi-minimum residual (TFQMR);

- Conjugate gradient squared (CGS).

Of these algorithms, GMRES is typically the most robust, but also has the highest memory requirement, since one additional vector is stored for each iteration. Transpose-free methods require matrix-vector multiplications only with $\mathbf{A}$ and not $\mathbf{A}^H$, which is useful for fast algorithms that are implemented in such a way that only matvecs with $\mathbf{A}$ can be computed.

The Krylov subspace associated with the linear system

$$\mathbf{Ax} = \mathbf{b} \tag{9.1}$$

is defined to be

$$K_k = \mathrm{span}\{\mathbf{b}, \mathbf{Ab}, \mathbf{A}^2\mathbf{b}, \ldots, \mathbf{A}^{k-1}\mathbf{b}\} \tag{9.2}$$

At the kth step, the CG algorithm finds the vector $\mathbf{x}_k$ in this subspace that minimizes the functional

$$f(\mathbf{x}) = \tfrac{1}{2}\mathbf{x}^T\mathbf{Ax} - \mathbf{b}^T\mathbf{x} \tag{9.3}$$

If the matrix $\mathbf{A}$ is SPD, this quadratic function has the property that its global minimum value is $f(\mathbf{A}^{-1}\mathbf{b})$, which can readily be seen by taking the vector derivative with respect to the argument $\mathbf{x}$. As the algorithm progresses, the Krylov subspace becomes larger. Because the CG algorithm minimizes $f(\mathbf{x})$ over these subspaces, the sequence of approximate solution vectors $\mathbf{x}_k$ approaches the solution to the linear system.

When the iteration count is equal to N (the size of the linear system), then as a consequence of the Cayley-Hamilton theorem the Krylov subspace K_N is equal to the full vector space R^N, and ignoring numerical error the vector $\mathbf{x}_N$ must be equal to $\mathbf{x} = \mathbf{A}^{-1}\mathbf{b}$. Because of this "finite completion" property, the CG algorithm can be viewed as a direct solution method when it is run for N iterations. Due to rounding error, however, the algorithm in practice fails to converge in N steps. For this reason, CG was not widely used in the early years after its development. As it became clear that CG is useful as an approximate iterative linear system solver when run for fewer then N iterations, and that despite the rounding error, the solution could be quite accurate for many problems at relatively low iteration counts ($k \ll N$), interest in CG was renewed and the algorithm is now widely used in applications of scientific computation.

9.1.1 CG Algorithm

The conjugate gradient algorithm is as follows:

Initialization ($k = 0$):

$$\mathbf{x}_0 \quad \text{(initial guess, almost always } \mathbf{x}_0 = 0) \tag{9.4a}$$

$$\mathbf{r}_0 = \mathbf{b} - \mathbf{A}\mathbf{x}_0 \quad \text{(initial residual vector)} \tag{9.4b}$$

$$\mathbf{d}_0 = \mathbf{r}_0 \quad \text{(initial search direction vector)} \tag{9.4c}$$

Loop over $k = 0, 1, 2, \ldots$

$$\gamma_k = \frac{\mathbf{r}_k^T \mathbf{r}_k}{\mathbf{r}_k^T \mathbf{A} \mathbf{d}_k} \tag{9.5a}$$

$$\mathbf{x}_{k+1} = \mathbf{x}_k + \gamma_k \mathbf{d}_k \quad \text{(update the approximation for } \mathbf{x}) \tag{9.5b}$$

$$\mathbf{r}_{k+1} = \mathbf{r}_k - \gamma_k \mathbf{A} \mathbf{d}_k \quad \text{(next residual vector)} \tag{9.5c}$$

$$\eta_k = \frac{\mathbf{r}_{k+1}^T \mathbf{r}_{k+1}}{\mathbf{r}_k^T \mathbf{r}_k} \tag{9.5d}$$

$$\mathbf{d}_{k+1} = \mathbf{r}_{k+1} + \eta_k \mathbf{d}_k \quad \text{(next search direction)} \tag{9.5e}$$

The vector $\mathbf{d}_k$ is referred to as the search direction, since $\mathbf{x}_{k+1}$ is obtained from the previous approximation $\mathbf{x}_k$ by adding a scalar multiple of $\mathbf{d}_k$. The vector $\mathbf{r}_k$ produced by the algorithm is equal to the residual error vector

$$\mathbf{r}_k = \mathbf{b} - \mathbf{A}\mathbf{x}_k \tag{9.6}$$

which is a measure of how close $\mathbf{x}_k$ is to the solution to the linear system.

9.1.2 CGNE and CGNR

For non-SPD matrices, the CG algorithm cannot be used directly. If $\mathbf{A}$ is not SPD, then we can solve the normal equation

$$\mathbf{A}^H \mathbf{A}\mathbf{x} = \mathbf{A}^H \mathbf{b} \tag{9.7}$$

The resulting algorithm is conjugate gradient on the normal equation (CGNE). Another possibility is to apply CG to the equation

$$AA^H y = b \qquad (9.8)$$

and then compute $x = A^H y$. This is known as CGNR.

The CGNE algorithm is as follows:

Initialization ($k = 0$):

$$r_0 = A^H b - A^H A x_0 \qquad (9.9a)$$
$$d_0 = r_0 \qquad (9.9b)$$

Loop over $k = 0, 1, 2, \ldots$

$$y = A d_k \qquad \text{(matvec by } A\text{)} \qquad (9.10a)$$

$$\gamma_k = \frac{r_k^H r_k}{y^H y} \qquad (9.10b)$$

$$x_{k+1} = x_k + \gamma_k d_k \qquad (9.10c)$$

$$y \leftarrow A^H y \qquad \text{(matvec by } A^H\text{)} \qquad (9.10d)$$

$$r_{k+1} = r_k - \gamma_k y \qquad (9.10e)$$

$$\eta_k = \frac{r_{k+1}^H r_{k+1}}{r_k^H r_k} \qquad (9.10f)$$

$$d_{k+1} = r_{k+1} + \eta_k d_k \qquad (9.10g)$$

9.1.3 Residual Error

Since we do not know the solution x, the error $e_k = x - x_k$ is not available. The norm of the residual error (9.6) is readily available during the iteration, and can be used in place of the actual solution error as an indicator of convergence of the algorithm. The relative residual error norm is

$$r_k = \frac{\|r_k\|}{\|r_0\|} \qquad (9.11)$$

Although the residual vector is not equal to the solution error e_k, the residual norm is equal to the solution error in a weighted norm. If we define

$$\|y\|_A = \|A y\| \qquad (9.12)$$

then the norm of the residual error is the A-norm of the solution error,

$$\|r_k\| = \|A(x - x_k)\| = \|e_k\|_A \qquad (9.13)$$

The residual error norm can be used to bound the norm of the solution error, since

$$
\begin{aligned}
\| \mathbf{r}_k \| &= \| \mathbf{A} \mathbf{e}_k \| \\
&\leq \| \mathbf{A} \| \| \mathbf{e}_k \| \\
&= | \lambda_{\max} | \| \mathbf{e}_k \|
\end{aligned}
\tag{9.14}
$$

where $\lambda_{\max}$ is the largest eigenvalue of $\mathbf{A}$. Since $\mathbf{A}$ is SPD, the eigenvalues are positive and equal to the singular values. Similarly,

$$
\begin{aligned}
\| \mathbf{e}_k \| &= \| \mathbf{A}^{-1} \mathbf{r}_k \| \\
&\leq \| \mathbf{A}^{-1} \| \| \mathbf{r}_k \| \\
&= \frac{1}{| \lambda_{\min} |} \| \mathbf{r}_k \|
\end{aligned}
\tag{9.15}
$$

Combining the two inequalities leads to

$$
\frac{1}{| \lambda_{\max} |} \| \mathbf{r}_k \| \leq \| \mathbf{e}_k \| \leq \frac{1}{| \lambda_{\min} |} \| \mathbf{r}_k \|
\tag{9.16}
$$

so that the residual error norm bounds the solution error.

9.1.4 Condition Number

For a normal matrix, the ratio

$$
\kappa(\mathbf{A}) = \frac{| \lambda_{\max} |}{| \lambda_{\min} |}
\tag{9.17}
$$

is the condition number of the matrix in the L^2 norm. In general, the condition number is defined to be

$$
\kappa(\mathbf{A}) = \| \mathbf{A} \| \| \mathbf{A}^{-1} \|
\tag{9.18}
$$

where $\| \cdot \|$ denotes a matrix norm. With the L^2 norm, the condition number for an arbitrary matrix is

$$
\kappa(\mathbf{A}) = \frac{\sigma_{\max}}{\sigma_{\min}}
\tag{9.19}
$$

where $\sigma_{\max}$ and $\sigma_{\min}$ are the largest and smallest singular values. This quantity is a measure of how near $\mathbf{A}$ is to a singular operator. If $\mathbf{A}$ is the identity, then the condition number is one. If $\mathbf{A}$ is singular, the condition number is infinite. An operator with a large condition number is said to be ill-conditioned.

If the condition number of $\mathbf{A}$ is close to one, then the residual error is always close to the solution error. If the condition number is large, then residual error may not be a good measure of solution error. In some cases, the condition number can be large, but residual error is still a good indicator of solution error, because the residual and error vectors may not be close to the eigenvectors with very small or very large eigenvalues and the extreme limits in the inequality (9.16) are not reached. As we will see later, the condition number also has a strong influence on the convergence rate of CG and other Krylov subspace algorithms.

9.1.5 Other Krylov Subspace Methods

One problem with CGNE is that this algorithm may converge more slowly than CG, because $A^H A$ is more poorly conditioned than A:

$$\kappa(A^H A) \simeq \kappa(A)^2 \tag{9.20}$$

Although as we will see shortly this simplistic reasoning may not always hold, it is often desirable to use iterative algorithms that work directly with non-SPD matrices. These include biconjugate gradient (BCG), biconjugate gradient-stabilized (BCG-stab), generalized minimum residual (GMRES), quasi-minimum residual (QMR), transpose-free quasi-minimum residual (TFQMR), and conjugate gradient squared (CGS).

BCG is similar to CG, but can have erratic convergence for non-SPD matrices. Modifications of BCG such as BCG-stab seek to reduce the erratic convergence behavior. As noted above, the generalized minimum residual method (GMRES) is typically the most robust of the Krylov subspace iterations, but to compute the kth iteration, k vectors from previous iterations must be stored, as opposed to two or three for CG. Restarted GMRES, or GMRES(n), throws away all but n of these stored vectors, reducing the memory requirement but also slowing convergence.

9.2 ITERATION COUNT ESTIMATES

In this section, we develop estimates for the number of iterations required to solve a linear system to a given error tolerance. According to the classical theory of Krylov subspace iterations, as the iteration count increases, the relative residual error norm (9.11) tends to decay as $r_k \simeq \rho^k$, where ρ is referred to as the asymptotic convergence factor of the iteration. For an SPD matrix, approximating the spectrum as an interval on the positive real axis leads to the convergence factor estimate [6]

$$\rho \simeq \frac{\sqrt{\kappa(A)} - 1}{\sqrt{\kappa(A)} + 1} \tag{9.21}$$

For the method of moments with the CGNR and CGNE algorithms, the matrix that enters into (9.21) is $A = Z^H Z$. Assuming that $\kappa(A) \simeq \kappa(Z)^2$, for large condition numbers the number of iterations required to obtain a residual error of $r_K \leq \epsilon$ is

$$K \simeq \kappa(\mathbf{Z}) \frac{|\ln \epsilon|}{2} \qquad (9.22)$$

where ϵ is the residual error tolerance parameter, and provides a stopping condition for the iteration. The iteration count estimate for CGNR and CGNE is proportional to the condition number of the moment matrix.

For other iterative methods such as the biconjugate gradient method (BCG) or generalized minimum residual (GMRES), which can be applied to non-SPD matrices, the asymptotic convergence factor depends on the distribution of the spectrum of $\mathbf{Z}$ in the complex plane. If the spectrum is approximated by a disk not containing the origin, the convergence factor becomes [7]

$$\rho = \frac{\kappa(\mathbf{Z}) - 1}{\kappa(\mathbf{Z}) + 1} \qquad (9.23)$$

which leads to the same iteration count estimate (9.22) as CGNE.

In practice, actual convergence rates can deviate significantly from these estimates. If the right-hand side belongs to an invariant subspace, for example, then the effective condition number is reduced to that of the operator restricted to the subspace. For a strongly nonnormal matrix, convergence can be very slow, even though the matrix may have the same condition number as another normal matrix. As demonstrated in Section 5.1, however, the degree of nonnormality of the moment matrix for electromagnetic integral operators is typically weak and (9.22) is often a reasonable estimate of the number of iterations required to reach an accurate solution for the method of moments.

9.3 CONDITION NUMBER ESTIMATES

The eigenvalue estimates obtained for the EFIE and MFIE operators in earlier chapters were used to analyze MoM solution error, but the estimates can also be used to determine the moment matrix condition number. Together with the iteration count estimate (9.22), these results will determine the computational cost of solving the moment method linear system with an iterative algorithm.

9.3.1 Circular Cylinder, TM-EFIE

For the circular cylinder, it followed from (3.6) that for a regular discretization, the moment matrix has a complete system of eigenvectors and is therefore a normal matrix. The singular values of a normal matrix are equal to the magnitudes of the eigenvalues, so that we can obtain the condition number from the extremal eigenvalues.

The largest eigenvalue of the moment matrix arises from maximizing (3.2) over the modeled modes. As a function of q, $|J_q(k_0 a) H_q^{(2)}(k_0 a)|$ is oscillatory and increases in magnitude for $|q| < k_0 a$, and decays monotonically for $|q| > k_0 a$. The maximum value occurs at $|q| \simeq k_0 a$. Using expansions of the Bessel and Hankel functions [8, Equations 8.441 #3, 8.443, and 8.454], we arrive at the asymptotic expression

$$J_\nu(\nu) H_\nu^{(2)}(\nu) \simeq \frac{6^{2/3}(1 + j\sqrt{3})}{9\,\Gamma^2(2/3)} \nu^{-2/3} \quad \nu \to \infty \tag{9.24}$$

From this result, it can be seen that the largest eigenvalue of the moment matrix is

$$\lambda_{max} \simeq \frac{\eta 2\pi (1 + j\sqrt{3})}{6^{4/3}\,\Gamma^2(2/3)} (k_0 a)^{1/3} \tag{9.25}$$

where we have neglected the small eigenvalue shift due to discretization error. This eigenvalue corresponds to a surface wave mode with spatial frequency k_0 on the cylinder. The magnitude of the eigenvalue grows with the 1/3 power of the electrical size of the cylinder. In Section 9.3.3, we will see that the corresponding growth rate exponent for the strip is 1/2, since the surface wave mode is more strongly self-coupled for a flat scatterer.

The smallest eigenvalue of the moment matrix is more difficult to determine, due to the internal resonances associated with closed conducting bodies discussed in Section 6.2. Near an internal resonance, the eigenvalue associated with the resonant mode is small in magnitude, and the condition number of the moment matrix is large. Due to the imaginary part of the spectral error, the locations of the resonance are shifted with respect to the exact internal resonance frequencies by discretization error, so the peaks in the matrix condition number are shifted slightly from the exact resonance frequencies.

If no modes are near resonance, then the smallest eigenvalues of the moment matrix correspond to nonradiating modes with rapid spatial oscillation. The spectrum of the operator $\mathcal{L}$ has an accumulation point at the origin, due to the vanishing eigenvalues of eigenfunctions of increasingly large order. Employing a finite basis to discretize the EFIE leads to a cutoff of the spectrum near this accumulation point at the mesh Nyquist frequency (2.63). For the circular cylinder, this corresponds to the Fourier mode with order $|q| = N/2$. Applying the large-order expansion $J_\nu(x) H_\nu^{(2)}(x) \sim j(\pi|\nu|)^{-1} + O(\nu^{-3})$, $\nu \to \infty$ [8, Equation 8.452] to $\lambda_{N/2} = (\eta\pi k_0 a/2) J_{N/2}(k_0 a) H_{N/2}^{(2)}(k_0 a)$ leads to the result

$$\lambda_{min} \simeq \frac{j\eta}{n_\lambda} \tag{9.26}$$

for the highest-order eigenvalue of the moment matrix. For an irregular discretization with variable mesh element sizes, the minimum eigenvalue is determined by the smallest discretization length or the largest value of the mesh density n_λ.

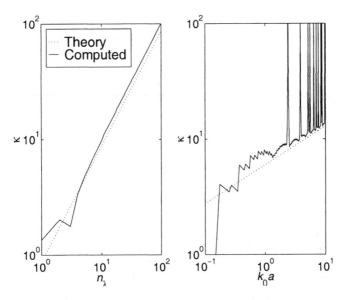

Figure 9.1: Condition number of the moment matrix for a circular cylinder. Integral equation: TM-EFIE. Discretization: point testing and pulse expansion functions. Left plot: fixed cylinder radius ($k_0 a = \pi$) as a function of mesh element density. Right plot: fixed mesh density as a function of cylinder radius. The dotted lines are the theoretical lower bound (9.27). (©2000 IEEE [9].)

As long as the highest-order eigenvalue (9.26) is smaller in magnitude than the eigenvalues of low-order modes that may be near to internal resonance, the moment matrix condition number can be approximated by the ratio of (9.25) to (9.26) as

$$\kappa(\mathbf{Z}) \geq 0.6\, n_\lambda (k_0 a)^{1/3} \tag{9.27}$$

Since internal resonances cause the magnitude of the smallest eigenvalue of the moment matrix to decrease, this estimate is a lower bound for the condition number. The estimate is compared to computed values in Figure 9.1.

If the kernel of the integral equation were smooth, then the high-order eigenvalues would fall off more rapidly, and the smallest eigenvalue of the moment matrix would be much smaller than (9.26). This would cause a much higher matrix condition number and increase the difficulty of solving the integral equation numerically. There is a trade-off between the difficulty of integrating the kernel of an integral operator of the form (3.5) and the matrix condition number. With a weakly singular kernel, integration is more difficult and aliasing error is generally large, but the moment matrix condition number is reasonable. For a smooth kernel, aliasing error is negligible, but the resulting moment matrix is ill-conditioned. For an operator such as the TE-EFIE that includes derivatives, as we will see shortly the high-order eigenvalues are large in magnitude. The intermediate case is a second-kind integral equation, for which the eigenvalues tend to a constant and conditioning is best.

9.3.2 Circular Cylinder, TE-EFIE

The spectrum of the TE-EFIE operator $\mathcal{N}$ has an accumulation point at $-j\infty$, since the eigenvalues of nonradiating modes increase in magnitude as the spatial frequency increases. The maximum eigenvalue of the moment matrix corresponds to the mode with order $q = N/2$ at the mesh Nyquist frequency and can be estimated as

$$\lambda_{max} \simeq -\frac{j\eta n_\lambda}{4} \tag{9.28}$$

where we have made use of $J_\nu'(x)H_\nu^{(2)\prime}(x) \sim -j|\nu|/(\pi x^2)$, $\nu \to \infty$. As with the TM polarization, internal resonances lead to small eigenvalues that dominate the condition number. Away from internal resonance frequencies, the smallest eigenvalues correspond to the surface wave mode with $|q| \simeq k_0 a$. For this mode, we can use $J_\nu'(\nu)H_\nu^{(2)\prime}(\nu) \simeq 0.2\,(1 - j\sqrt{3})\nu^{-4/3}$, $\nu \to \infty$ to obtain

$$\lambda_{min} \simeq 0.3\,\eta(1 - j\sqrt{3})(k_0 a)^{-1/3} \tag{9.29}$$

The resulting condition number estimate is

$$\kappa(\mathbf{Z}) \simeq 0.4\,n_\lambda(k_0 a)^{1/3} \tag{9.30}$$

which is of the same order as the TM result.

9.3.3 Flat Strip, TM-EFIE

As for the cylinder, employing a finite basis to discretize the EFIE leads to a cutoff of the spectrum near this accumulation point, so that the spectrum of $\mathbf{Z}$ corresponds to the N lowest-order eigenvalues of $\mathcal{L}$. From (5.15), the eigenvalue of the moment matrix with the smallest magnitude corresponds to the normalized spatial frequency $\beta_q = n_\lambda/2$. With (5.15) and neglecting spectral error, the smallest eigenvalue is the same as that obtained in (9.26) for the circular cylinder. Since a high-order mode does not radiate strongly, the mode self-coupling is local and the eigenvalue is not strongly sensitive to the global geometry of the scatterer.

As with the circular cylinder, the largest eigenvalue of the TM-EFIE operator corresponds to the surface wave mode. From (5.7), the eigenvalue is

$$\lambda_{max} \simeq \frac{\eta\sqrt{2}}{3}(1 + j)D^{1/2} \tag{9.31}$$

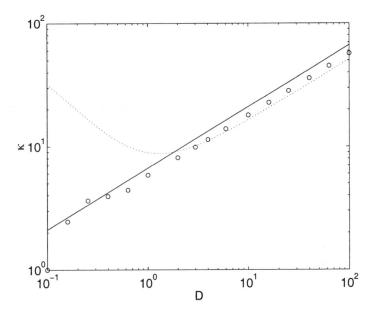

Figure 9.2: Condition number of the moment matrix for the flat strip as a function of the width D in wavelengths. Integral equation: TM-EFIE. Discretization: point testing and pulse expansion functions, $n_\lambda = 10$. Circles: computed moment matrix condition number. Solid line: theoretical approximation, (9.32). Dotted line: theoretical approximation, (5.6) and (5.7), including discretization error. (©2001 John Wiley & Sons [10].)

The surface wave mode can be considered to be antiresonant, since its eigenvalue grows with the width of the strip. If the strip lies in a lossy medium, so that k_0 has a nonzero imaginary part, then the singularities in the integrand of (5.13) at $|\beta| = 1$ are eliminated, and for large D the maximum eigenvalue becomes independent of the scatterer size.

Since the moment matrix is nonnormal for the flat strip, the singular values are larger than the eigenvalues, and the condition number is larger than the ratio of the maximum and minimum eigenvalues. The degree of nonnormality is weak, however, and the eigenvalue ratio provides a reasonable estimate of the condition number. By making use of the extremal eigenvalue estimates obtained above, we have

$$\kappa(\mathbf{Z}) \simeq \frac{2}{3} n_\lambda D^{1/2} \tag{9.32}$$

Figure 9.2 shows the computed condition number for $n_\lambda = 10$ as a function of D and the theoretical estimate with and without discretization error and higher-order terms in the eigenvalue estimates obtained in Section 5.1.

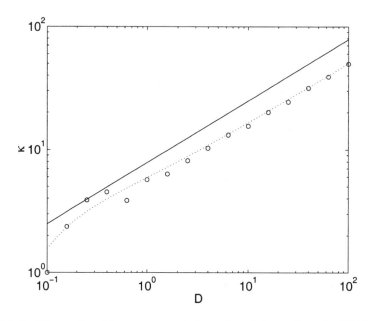

Figure 9.3: Condition number of the moment matrix for the flat strip. Integral equation: TE-EFIE. Discretization: point testing and pulse basis functions, $n_\lambda = 10$. Circles: computed moment matrix condition number. Solid line: theoretical approximation, (9.33). Dotted line: theoretical approximation, (5.39) and (5.40), including discretization error. (©2001 John Wiley & Sons [10].)

9.3.4 Flat Strip, TE-EFIE

For the TE polarization, as for the circular cylinder, the largest eigenvalue of the moment matrix arises from the discrete mode with largest spatial frequency. This corresponds to the estimate (5.39) evaluated at $q = N/2$, which leads to the same result as (9.28). The smallest eigenvalue arises from the surface wave mode and is approximated by (5.40). From the ratio of the extremal eigenvalues, the moment matrix condition number can be estimated as

$$\kappa \simeq \frac{\pi}{4} n_\lambda D^{1/2} \tag{9.33}$$

Figure 9.3 shows the condition number as a function of the strip length for a discretization density of $n_\lambda = 10$.

9.3.5 Parallel Strip Resonator

The eigenvalues of resonant modes for a parallel strip resonator were analyzed in Section 6.3. In order to obtain condition number estimates, we also require estimates of the largest eigenvalues of the EFIE operator. For the parallel strip resonator, surface wave modes radiated parallel to the strips and are not strongly coupled from one strip to the other, so the single strip surface wave mode eigenvalue estimates can be used to estimate the largest eigenvalue of the TM-EFIE for the cavity. For the TE-EFIE, the largest eigenvalue is determined by the mode at the mesh Nyquist frequency. Since this mode is nonradiating and only locally coupled, the estimate (9.28) for the circular cylinder can be used for the cavity.

For large cavity dimensions, there are many closely spaced resonances, so we can assume that regardless of the dimensions or frequency one mode is near enough to resonance that the imaginary part of the eigenvalue can be taken to be zero, and the real part (6.19) can be employed as an estimate of the magnitude of λ_{min}. For the TM polarization, this leads to a condition number estimate of

$$\kappa^{TM} \simeq \frac{4}{3} \frac{D^{3/2}}{\alpha\sqrt{W}} \qquad (9.34)$$

where we have retained only the leading-order term of (6.19). D is the cavity depth in wavelengths and α is the fractional part of twice the cavity width W in wavelengths. For the TE polarization,

$$\kappa^{TE} \simeq \frac{n_\lambda D}{2\alpha\sqrt{W}} \qquad (9.35)$$

The growth rate for the TE polarization with D is not as large as that of the TM case, since the largest eigenvalue of $\mathcal{N}$ does not depend on electrical size, whereas the largest eigenvalue of $\mathcal{L}$ increases in magnitude with electrical size.

From (9.34) and (9.35), it can be seen that the condition number of the moment matrix is maximal at the smallest possible value of α. This corresponds to the resonance of the $q = 1$ current mode, for which the fields are approximated by the TM_{1n} rectangular waveguide mode. The normalized spatial frequency of the mode is $\beta_1 = 1/(2D)$, and at resonance the smallest eigenvalue is equal to the real part given by (6.21). With (9.31), the maximum condition number is

$$\kappa_{max}^{TM} \simeq \frac{16}{3} \frac{D^{7/2}}{W^{3/2}} \qquad (9.36)$$

for the TM polarization. For the TE polarization,

$$\kappa_{max}^{TE} \simeq \frac{2n_\lambda D^3}{W^{3/2}} \qquad (9.37)$$

As shown by these estimates, the condition number of the moment matrix grows rapidly with the cavity depth D. Since the parameter α increases with the mode number q, the most extreme ill-conditioning is confined to narrow bands near the lowest-order resonances of the cavity corresponding to small values of q, where the cavity width is near to an integer or half-integer number of wavelengths.

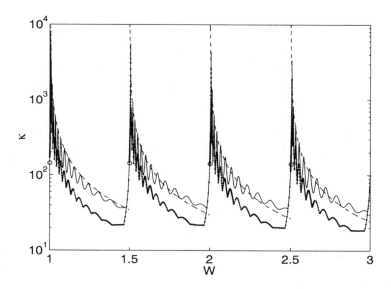

Figure 9.4: Moment matrix condition number for a parallel strip resonator with depth $D = 10$ in wavelengths, as a function of width in wavelengths. Integral equation: TM-EFIE. Discretization: point testing and pulse expansion functions, $n_\lambda = 10$. Solid line/dots: computed value, open cavity. Solid line: computed value, half-open cavity. Dashed line: theoretical estimate, (9.34). Circles: (6.19) and (6.24), including discretization error (6.25), at exact resonances of the lowest-order mode ($q = 1$). (©1999 John Wiley & Sons [11].)

These theoretical condition number estimates for the cavity can be validated by comparison with numerical results. Figure 9.4 shows the condition number of the moment matrix for a cavity of length $D = 10$ in wavelengths as a function of width. The more precise theoretical estimate obtained by taking into account both (6.19) and (6.24), as well as the spectral error (6.25), is shown at the locations of the resonances of the TM_{1n} mode. As W increases, the cavity mode eigenvalues move past the origin in the complex plane, and the condition number is largest when one of the modes is at resonance and its eigenvalue has a vanishing imaginary part. The smaller peaks correspond to higher-order modes moving through resonance, and the largest peaks near integer or half-integer widths correspond to resonances of the $q = 1$ mode, which is most strongly confined to the cavity and has the highest quality factor.

Figure 9.5 is an expansion of a small interval of the results shown in Figure 9.4. The theoretical locations of the resonances of the EFIE are marked by circles. The shifting of the peaks of the computed condition numbers away from these locations is caused by discretization error.

Numerical results are shown for both the open and half-open parallel strip res-

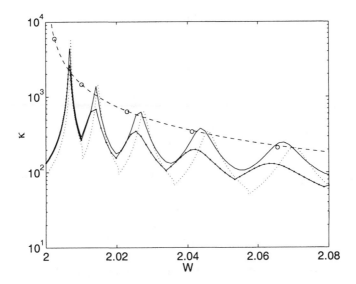

Figure 9.5: Condition number of moment matrix for the same cavity as in Figure 9.4, over a smaller range of the width W. Solid line/dots: computed value, open cavity. Solid line: computed value, half-open cavity. Dashed line: theoretical estimate, (9.34). Dotted line: (6.19) and (6.24), including discretization error (6.25). Circles: (6.19) and (6.24), without discretization error, at exact resonances. (©1999 John Wiley & Sons [11].)

onators. Away from resonant peaks, the condition number for the half-open cavity is approximately twice that of the open cavity.

9.3.6 Higher-Order Basis Functions

It is known that the condition number of the moment matrix increases rapidly with the polynomial order p of the basis, at least as fast as p^2 [12]. The spectral estimates obtained in Section 5.1 are only valid for modes numbers in the range $|q| \le N/2$, where N is the number of mesh elements. This suffices for the low-order basis functions considered above, but for higher-order basis functions, the additional degrees of freedom on each mesh element mean that the order of the mode with smallest eigenvalue is greater than $N/2$. A more sophisticated treatment is required in order to obtain a condition number estimate for higher-order basis functions.

Let $\mathbf{U}$ be the unitary transformation with matrix elements

$$U_{nq} = N^{-1/2} e^{-j\beta_q k_0 x_n} \tag{9.38}$$

where β_q and x_n are defined as before. Transforming each block of the moment matrix corresponding to the polynomial functions indexed by a and b leads to

$$A_{qr,ab} = \sum_{m,n=1}^{N} U_{mq}^* Z_{mn,ab} U_{nr} \tag{9.39}$$

By making use of (8.9), this matrix element can be written as

$$A_{qr,ab} = \frac{\eta}{2n_\lambda^2 D} \int_{-\infty}^{\infty} \frac{d\beta}{\sqrt{1-\beta^2}} \frac{\sin\left[\pi D(\beta - \beta_q)\right]}{\sin\left[\pi(\beta - \beta_q)/n_\lambda\right]} \frac{\sin\left[\pi D(\beta - \beta_r)\right]}{\sin\left[\pi(\beta - \beta_r)/n_\lambda\right]} F_a(-\beta)F_b(\beta) \tag{9.40}$$

Expanding the integrand about the maxima for the diagonal elements with $q = r$ leads to

$$A_{qq,ab} \simeq \frac{\eta}{2} \sum_{s=-\infty}^{\infty} \frac{F_a(-\beta_{q,s})F_b(\beta_{q,s})}{\sqrt{1-\beta_{q,s}^2}} \tag{9.41}$$

where $\beta_{q,s} = \beta_q + sn_\lambda$.

The minimum eigenvalue of the moment matrix can be estimated from the smallest eigenvalue of $A_{qq,ab}$ for the largest value of q, which is $q = N/2$. For high-order modes, the EFIE operator approaches its static limit and is approximately self-adjoint, and the matrix $\mathbf{A}_{qq}$ with elements given by $A_{qq,ab}$ for $q = N/2$ is approximately Hermitian. We can therefore employ the variational bound

$$\left|\hat{\lambda}_{\min}\right| \leq \left|\frac{\mathbf{v}^H \mathbf{A}_{N/2,N/2} \mathbf{v}}{\mathbf{v}^H \mathbf{v}}\right| \tag{9.42}$$

to estimate the smallest eigenvalue. It turns out that the minimum eigenvalue is obtained with the trial vector with elements given by $v_a = F_a(k)$, which leads to

$$\left|\hat{\lambda}_{\min}\right| \leq \left|\frac{\eta}{2F^{(p)}(\beta,\beta)} \sum_s \frac{F^{(p)^2}(\beta,\beta_{q,s})}{\sqrt{1-\beta_{q,s}^2}}\right| \tag{9.43}$$

By the variational expression (9.42), minimizing this expression over β leads to an upper bound for the smallest eigenvalue of the moment matrix.

We will now apply this eigenvalue bound to the Legendre polynomial expansion. Using the asymptotic approximation $j_a(x) \sim \cos\left[x + (a + 1/2)^2/(2x) - \pi(a-1)/2\right]/x$ for the spherical Bessel function, it can be shown that the minimum of the right-hand side of (9.43) occurs for $\beta = n_\lambda/2 + n_\lambda s'$, where s' is a large integer. For β of this form, $F^{(p)}(\beta,\beta) \simeq (p+1)(p+2)/(2\beta^2)$, and

$$F^{(p)}(\beta,\beta_{q,s}) \simeq (\beta\beta_{q,s})^{-1} \sum_{a=0}^{p/2} (4a+1) \cos\left[\frac{(2a+1/2)^2}{2\pi s}\right] \tag{9.44}$$

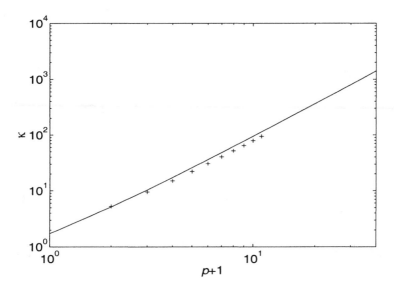

Figure 9.6: Moment matrix condition number as a function of polynomial order for Legendre polynomial expansion applied to a flat strip, $D = 1$, $n_\lambda = 4$. Integral equation: TM-EFIE. Solid line: theoretical estimate, (9.46). Pluses: numerical results.

Using this in (9.43) leads to the estimate

$$\left|\hat{\lambda}_{\min}\right| \simeq \frac{\eta}{n_\lambda(p+1)(p+2)} \sum_{s=p/2}^{\infty} (2/\pi^2)s^{-3} \left\{ \sum_{a=0}^{p/2} (4a+1) \cos\left[\frac{(2a+1/2)^2}{2\pi s}\right] \right\}^2 \quad (9.45)$$

for p even, with a similar result for p odd. The summation over q can be evaluated numerically and is nearly independent of p, with an approximate value of 3.1. The resulting condition number estimate is

$$\kappa \simeq 0.2\, n_\lambda(p+1)(p+2)D^{1/2} \quad (9.46)$$

This estimate is compared to numerical results for the Legendre expansion applied to a flat strip of length $D = 1$ in wavelengths in Figure 9.6.

9.3.7 Flat Plate—3D

For a flat plate, the extremal eigenvalues of the moment matrix for the EFIE correspond to the largest curl-free eigenvalue (7.9) and the smallest divergence-free eigenvalue

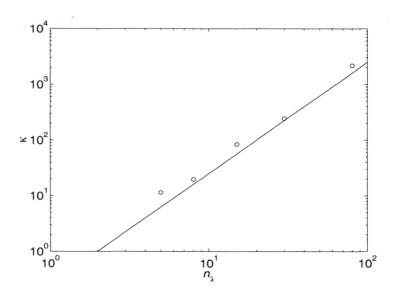

Figure 9.7: Moment matrix condition number for a square plate. Integral equation: EFIE. Solid line: theoretical estimate, (9.47). Circles: computed.

(7.10). These eigenvalue estimates lead to the condition number estimate

$$\kappa \simeq \frac{\pi^2}{k_0^2 h^2} = \frac{n_\lambda^2}{4} \tag{9.47}$$

This result breaks down as the plate size d becomes large relative to the wavelength $\lambda = 2\pi/k_0$, since the eigenvalues of the curl-free and divergence-free surface wave modes depend on $k_0 d$ and eventually will surpass the eigenvalues of the nonradiating modes. For a large enough plate dimension or frequency, the condition number begins to grow with the electrical size of the scatterer.

Since the high-spatial frequency, nonradiating modes that determine the condition number estimate (9.47) radiate evanescent fields and are not globally coupled, the eigenvalues depend only on local properties of the scatterer. Therefore, unless resonance or large electrical size leads to eigenvalues that are larger than (7.9) or smaller than (7.10), the estimate (9.47) holds for arbitrary scatterers.

Figure 9.7 compares the theoretical condition number estimate to numerical results for a square plate. For the computed values, RWG basis functions are employed on a regular triangular mesh. The plate size is $d = 1\lambda$ for the smaller values of n_λ, and $d = 0.1\lambda$ for the largest.

9.4 LOW-FREQUENCY BREAKDOWN

For a fixed mesh, the estimate (9.47) shows that the condition number of the moment matrix for the EFIE grows as $1/k_0^2$ as the frequency decreases. This is observed in practice as the low-frequency breakdown of the method of moments. As the frequency decreases, the curl-free and divergence-free modes discussed in Section 7.1 decouple and the eigenvalues separate into two groups. If the basis used to discretize the EFIE is such that the divergence-free modes can be separated from the curl-free modes, then the two subspaces can be rescaled to improve the conditioning of the discretized operator at low frequencies. This can be accomplished using the loop-star or loop-tree techniques [13–16]. In this section, we analyze low-frequency breakdown and approaches to overcoming this problem for the EFIE.

We have already obtained estimates for the largest curl-free or TE type mode eigenvalue and the smallest divergence-free or TM type eigenvalue. To analyze low-frequency breakdown, we need to estimate the smallest curl-free eigenvalue and the largest divergence-free eigenvalue. For an electrically small flat PEC plate, since the DC or constant current mode is divergence-free, the smallest curl-free eigenvalue corresponds to a mode with the lowest possible nonzero spatial frequency, which is $\beta = \pi/(k_0 d)$. For this mode, (7.6) yields the estimate

$$\lambda_{\min}^{\text{TE}} \simeq -\frac{j\eta}{2} \frac{\pi}{k_0 d} \tag{9.48}$$

for this eigenvalue. The curl-free part of the spectrum of the moment matrix lies between the two extremal eigenvalues given by (7.9) and (9.48).

For small $k_0 h$, the largest divergence-free eigenvalue corresponds to the mode with lowest spatial frequency ($\beta \simeq 0$). If $k_0 d \ll 1$, the approximation used in obtaining (7.4) breaks down for $\beta = 0$, since k_z is rapidly varying near the maxima of the sinc functions at $k_x = 0$, $k_y = 0$. A more accurate evaluation of the integral in (7.3) is obtained by expanding the sinc functions, retaining terms up to second order, and integrating up to the first zeros of the sinc functions. This leads to the estimate

$$\lambda_{\max}^{\text{TM}} \simeq \frac{k_0 \eta d^2}{2\pi^2} \int_0^{2\pi/d} \int_0^{2\pi/d} \frac{dk_x \, dk_y}{k_z} \left[1 - \frac{k_x^2 d^2}{24} - \frac{k_y^2 d^2}{24} \right] \tag{9.49}$$

Evaluating the integral in the limit as $k_0 \to 0$ yields

$$\lambda_{\max}^{\text{TM}} \simeq \frac{j\eta k_0 d}{2\pi} \left[(4 - \pi^2/9) \log(1 + \sqrt{2}) - \pi^2 \sqrt{2}/9 \right] \simeq \frac{j\eta}{2} \frac{k_0 d}{\pi} \tag{9.50}$$

At low frequencies, the divergence-free part of the spectrum lies between the values given by (7.10) and (9.50).

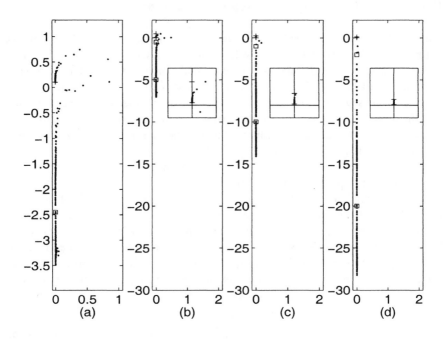

Figure 9.8: Moment matrix spectra for 1 m × 1 m plate for decreasing frequencies. (a) f = 300 MHz, (b) f = 150 MHz, (c) f = 75 MHz, and (d) f = 37.5 MHz. The spectra are normalized to η = 1. Squares: theoretical extremal curl-free eigenvalue estimates, (7.9) and (9.48). Pluses: theoretical extremal divergence-free eigenvalue estimates, (7.10) and (9.50). As the frequency becomes small, the two parts of the spectrum separate. The curl-free eigenvalues move to $-j\infty$, and the divergence-free eigenvalues move towards the origin. Magnified views of the eigenvalues near the origin are shown as insets.

To validate these estimates, Figure 9.8 shows the moment matrix spectrum for a plate of side d = 1 m at various frequencies. RWG vector basis functions are employed on a regular triangular mesh. The discretization length h is taken to be the minimum distance between mesh nodes, 0.1 m. As discussed in Chapter 7, for large $k_0 d$, the spectrum lies on a question mark shaped arc in the complex plane, which extends from the negative imaginary axis, through $\eta/2$ on the real axis, where the curl-free and divergence-free parts of the spectrum join, and ending near the origin along the positive imaginary axis. This can be seen in Figure 9.8(a). As k_0 decreases, the two parts of the spectrum separate, moving towards the accumulation points of the spectrum of $\mathcal{T}$ at 0 and $-j\infty$. This sequence of spectra illustrates the phenomenon of low-frequency breakdown.

9.4.1 Helmholtz Decomposition

To overcome the ill-conditioning caused by low-frequency breakdown, a vector basis is required that allows the curl-free and divergence-free modes to be separated. This is referred to as a Helmholtz decomposition.

To analyze the improvement in condition number with a Helmholtz decomposition, we will first assume that the decomposition is exact, so that the curl-free and divergence-free modes can be explicitly separated. If the divergence-free moment matrix block is scaled by $1/(k_0 h)$, and the curl-free block by $k_0 h$, then the condition number becomes

$$\kappa \simeq \frac{d^2}{\pi^2 h^2} \tag{9.51}$$

which is independent of frequency and low-frequency breakdown no longer occurs.

The eigenvalue estimates obtained above can be used to further optimize the condition number. The optimal scaling factors are $k_0 d/\pi$ and $-\pi/(k_0 h)$, in which case the spectrum of the scaled operator is a single interval in the complex plane with extrema at $-j$ and $-jd/h$. The condition number improves to

$$\kappa \simeq d/h \tag{9.52}$$

which is the same as that of the discretized integral equation for the Laplace or static problem ($k_0 \to 0$). This is to be expected, because at very low frequencies the dynamic problem decouples into two static problems that could be solved individually.

For the loop-star and loop-tree bases, the loop space is divergence-free, and the restriction of the discretized operator to this space is well conditioned, with $\kappa \simeq d/h$. The restriction to the star or tree spaces, however, has been observed empirically to be poorly conditioned [17, 18], so that the estimates (9.51) and (9.52) are not achieved by the loop-star or loop-tree approaches.

An analysis of the matrix spectrum shows that the ill-conditioning of the tree part of the moment matrix is due to the existence of modes in the discrete tree space with eigenvalues that are much smaller than eigenvalues associated with a continuous curl-free space. Figure 9.9(a) shows the eigenfunction corresponding to the eigenvalue with smallest magnitude of the tree space matrix for a square plate. The charge accumulation for this mode is small, since the current flows back and forth between the vertical cuts that define the tree space. Figure 9.9(b) shows the approximate curl-free mode with smallest eigenvalue for the RWG discretization of the same problem, which corresponds to the eigenvalue with smallest positive imaginary part in Figure 9.8(d). Strong charge accumulations exist at the corners of the plate.

The problem with the loop-star and loop-tree methods is that while the mode shown in 9.9(a) does have a nonzero divergence, the divergence is too small as compared to the curl-free modes that would be realized with an exact Helmholtz decomposition. For this example, the ratio of the smallest curl-free eigenvalue to the smallest

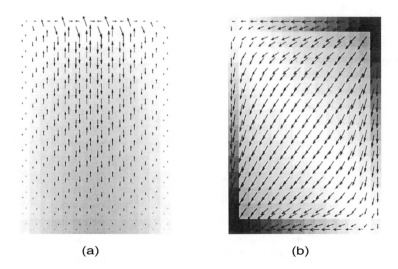

(a)   (b)

Figure 9.9: (a) Mode corresponding to smallest eigenvalue of the tree space interaction matrix for a square plate of side 0.025λ. (b) Mode corresponding to smallest curl-free eigenvalue. Superimposed on the vector fields are the charge densities associated with each mode. The curl-free mode is associated with a relatively strong charge density or divergence, whereas the tree space mode is nearly divergence-free, so the eigenvalue encroaches on the divergence-free part of the spectrum and causes ill-conditioning.

tree space eigenvalue is approximately 30. The condition number of the tree space matrix is roughly 30 times larger than that of the curl-free restriction of the full moment matrix, due to the presence of low-charge or nearly divergence-free modes in the tree space.

One way to improve this residual ill-conditioning is to multiply the tree space matrix with the inverse of a discretization of the divergence operator [17, 18]. Since the low-charge modes have small divergence, multiplication by this inverse operator increases the eigenvalues of these modes relative to other modes and compresses the spectrum of the tree matrix.

9.5 PRECONDITIONERS

Since the condition number of a matrix strongly influences the convergence of an iterative solution algorithm, if the matrix in a linear system can be effectively transformed so that the matrix has a smaller condition number, the computational cost of solving the linear system can be improved. A left preconditioner $\mathbf{M}$ transforms the linear system to

$$\mathbf{M}^{-1}\mathbf{A}\mathbf{x} = \mathbf{M}^{-1}\mathbf{b} \tag{9.53}$$

If an iterative algorithm is applied to the preconditioned linear system, the condition number that governs convergence is that of $\mathbf{M}^{-1}\mathbf{A}$, which should satisfy

$$\kappa(\mathbf{M}^{-1}\mathbf{A}) \ll \kappa(\mathbf{A}) \tag{9.54}$$

In implementing a preconditioned iteration, it would be computationally costly to actually compute $\mathbf{M}^{-1}$ or its product with $\mathbf{A}$. An iterative algorithm can be modified to include a preconditioner by solving

$$\mathbf{M}\mathbf{y} = \mathbf{c} \tag{9.55}$$

for $\mathbf{y}$ at each iteration. If $\mathbf{M}$ were equal to $\mathbf{A}$, this would take as much work as solving the original linear system, but the condition number of the matrix in (9.54) would be unity and the iteration would only require one step. Thus, we want the preconditioning matrix to be similar enough to $\mathbf{A}$ that the preconditioned matrix has a condition number as close to unity as possible, but structured in a way that the linear system (9.55) can be solved efficiently. The preconditioner can either be constructed blindly from the matrix $\mathbf{A}$, or the physics of the problem that led to $\mathbf{A}$ can be used to suggest a form for $\mathbf{M}$.

A common preconditioner for the method of moments is a nearest neighbor matrix of interactions between closely spaced testing and expansion functions. For 2D scattering, this leads to a banded matrix. In 3D, the matrix is not banded, but is typically sparse, so an algorithm for solving a sparse linear system exactly or approximately must be used to solve (9.55). The physical radius of the near-neighbor region determines the effectiveness of the preconditioner. The larger the radius, the harder it is to solve (9.55) at each step in the iteration, but convergence of the iteration improves.

As shown in the previous section, condition number increases with the mesh density due to eigenvalues associated with high-order, nonradiating modes. For a fixed physical radius, the nearest neighbor preconditioner overcomes the growth of the condition number with the discretization density, because the preconditioner accurately models the localized self-interaction of evanescent modes with high spatial frequency. As the mesh density increases, the number of matrix elements falling within the near neighbor radius increases, and (9.55) becomes harder to solve. Since the preconditioner does not model the long-range interactions that determine the surface wave eigenvalues, the condition number can still increase with electrical size of the scatterer. Near-neighbor preconditioners also do not overcome ill-conditioning due to resonant effects, since resonance is produced by global interactions.

Other physics-based preconditioners can be developed from virtually any scattering approximation that can be efficiently computed. For rough surfaces, a preconditioner can be constructed using a flat approximating surface [19]. High-frequency approximations for long-range interactions can also be used to develop preconditioners. Calderon identities can be used to precondition the EFIE, due to the approximate inverse relationship between the TM-EFIE and TE-EFIE operators [20, 21].

9.6 SUMMARY

We have seen that the convergence behaviors of Krylov subspace type iterative linear system solution algorithms are strongly influenced by the moment matrix condition number. Condition number estimates can be obtained from the spectral estimates developed in earlier chapters. Factors that increase the matrix condition number and slow the convergence of iterative algorithms include mesh refinement, which leads to a linear increase in condition number for 2D problems and a quadratic increase for 3D problems, resonance, and increasing polynomial order of basis functions. For the EFIE, decreasing the frequency effectively increases the mesh element density per wavelength, also causing ill-conditioning. A spectral analysis of low-frequency breakdown provides insight into this effect and guidance for optimizing low-frequency moment method approaches.

REFERENCES

[1] C. C. Lu and W. C. Chew, "A multilevel algorithm for solving a boundary integral equation of scattering," *Micro. Opt. Tech. Lett.*, vol. 7, pp. 466–470, July 1994.

[2] L. Tsang, C. H. Chan, and H. Sangani, "Application of a banded matrix iterative approach to Monte Carlo simulations of scattering of waves by random rough surface: TM case," *Microw. Opt. Tech. Lett.*, vol. 6, pp. 148–151, Feb. 1993.

[3] D. A. Kapp and G. S. Brown, "A new numerical method for rough-surface scattering calculations," *IEEE Trans. Ant. Propag.*, vol. 44, pp. 711–721, May 1996.

[4] G. H. Golub and H. A. van der Vorst, "Closer to the solution: Iterative linear solvers," in *The State of the Art in Numerical Analysis* (I. S. Duff and G. A. Watson, eds.), pp. 63–92, Oxford: Clarendon Press, 1997.

[5] J. C. West and J. M. Sturm, "On iterative approaches for electromagnetic rough-surface scattering problems," *IEEE Trans. Ant. Propag.*, vol. 47, pp. 1281–1288, Aug. 1999.

[6] G. H. Golub and C. F. V. Loan, *Matrix Computations*. Baltimore: Johns Hopkins University Press, 2 ed., 1993.

[7] T. A. Driscoll, K.-C. Toh, and L. N. Trefethen, "From potential theory to matrix iterations in six steps," *SIAM Review*, vol. 40, pp. 547–578, 1998.

[8] I. S. Gradshteyn and I. M. Ryzhik, *Table of Integrals, Series, and Products*. San Diego: Academic Press, 5 ed., 1994.

[9] K. F. Warnick and W. C. Chew, "Accuracy of the method of moments for scattering by a cylinder," *IEEE Trans. Micr. Th. Tech.*, vol. 48, pp. 1652–1660, Oct. 2000.

[10] K. F. Warnick and W. C. Chew, "On the spectrum of the electric field integral equation and the convergence of the moment method," *Int. J. Numer. Meth. Engr.*, vol. 51, pp. 31–56, May 2001.

[11] K. F. Warnick and W. C. Chew, "Convergence of moment method solutions of the electric field integral equation for a 2D open cavity," *Microw. Opt. Tech. Lett.*, vol. 23, pp. 212–218, Nov. 1999.

[12] M. Maischak, P. Mund, and E. P. Stephan, "Adaptive multilevel BEM for acoustic scattering," *Comp. Meth. Appl. Mech. Engrg.*, vol. 150, pp. 351–367, 1997.

[13] J. R. Mautz and R. F. Harrington, "An E-field solution for a conducting surface small or comparable to the wavelength," *IEEE Trans. Ant. Propag.*, vol. 32, pp. 330–339, Apr. 1984.

[14] E. Arvas, R. F. Harrington, and J. R. Mautz, "Radiation and scattering from electrically small conducting bodies of arbitrary shape," *IEEE Trans. Ant. Propag.*, vol. 34, pp. 66–77, Jan. 1986.

[15] M. Burton and S. Kashyap, "A study of a recent, moment-method algorithm that is accurate to very low frequencies," *Appl. Comp. Electromag. Soc. J.*, vol. 10, pp. 58–68, Nov. 1995.

[16] W. Wu, A. W. Glisson, and D. Kajfez, "A study of two numerical solution procedures for the electric field integral equation at low frequency," *Appl. Comp. Electromag. Soc. J.*, vol. 10, pp. 69–80, Nov. 1995.

[17] W. C. Chew, J. S. Zhao, and J. M. Song, "Solving Maxwell's equations from zero to microwave frequencies," *30th Plasmadynamics and Lasers Conference*, Paper 99-3729, Norfolk, VA, American Institute of Aeronautics and Astronautics, June 28–July 1, 1999.

[18] J. S. Zhao and W. C. Chew, "Integral equation solution of Maxwell's equations from zero frequency to microwave frequencies," *IEEE Trans. Ant. Propag.*, vol. 48, pp. 1635–1645, Oct. 2000.

[19] L. Tsang, C. H. Chan, and K. Pak, "Monte Carlo simulation of a two-dimensional random rough surface using the sparse-matrix flat-surface iterative approach," *Electron. Lett.*, vol. 29, pp. 1153–1154, June 1993.

[20] S. Amini and S. M. Kirkup, "Solution of Helmholtz equation in the exterior domain by elementary boundary integral methods," *J. Comp. Phys.*, vol. 118, pp. 208–221, 1995.

[21] G. C. Hsiao and R. E. Kleinman, "Mathematical foundations for error estimation in numerical solutions of integral equation in electromagnetics," *IEEE Trans. Ant. Propag.*, vol. 45, pp. 316–328, Mar. 1997.

About the Author

Karl F. Warnick received a B.S. magna cum laude with university honors in 1994 and a Ph.D. in 1997, both from Brigham Young University in Provo, Utah. From 1998 to 2000, he was a postdoctoral research associate and visiting assistant professor in the Center for Computational Electromagnetics at the University of Illinois at Urbana–Champaign. In 2000, he joined the Department of Electrical and Computer Engineering at Brigham Young University, where he is currently an associate professor. Professor Warnick was a recipient of the National Science Foundation Graduate Research Fellowship, Outstanding Faculty Member award for Electrical and Computer Engineering in 2005, and the BYU Young Scholar Award (2007), and he served as the technical program cochair for the 2007 IEEE International Symposium on Antennas and Propagation. He has published more than 100 conference articles and scientific journal papers and is a coauthor of the book *Problem Solving in Electromagnetics, Microwave Circuits, and Antenna Design for Communications Engineering* (Artech House, 2006) with Peter Russer.

Index

Grid Computing for Electromagnetics, Luciano Tarricone and Alessandra Esposito

Iterative and Self-Adaptive Finite-Elements in Electromagnetic Modeling, Magdalena Salazar-Palma, et al.

Numerical Analysis for Electromagnetic Integral Equations, Karl F. Warnick

Parallel Finite-Difference Time-Domain Method, Wenhua Yu, et al.

Quick Finite Elements for Electromagnetic Waves, Giuseppe Pelosi, Roberto Coccioli, and Stefano Selleri

Understanding Electromagnetic Scattering Using the Moment Method: A Practical Approach, Randy Bancroft

Wavelet Applications in Engineering Electromagnetics, Tapan K. Sarkar, Magdalena Salazar-Palma, and Michael C. Wicks

For further information on these and other Artech House titles, including previously considered out-of-print books now available through our In-Print-Forever® (IPF®) program, contact:

Artech House	Artech House
685 Canton Street	46 Gillingham Street
Norwood, MA 02062	London SW1V 1AH UK
Phone: 781-769-9750	Phone: +44 (0)20 7596-8750
Fax: 781-769-6334	Fax: +44 (0)20 7630 0166
e-mail: artech@artechhouse.com	e-mail: artech-uk@artechhouse.com

Find us on the World Wide Web at: www.artechhouse.com